Copyright © 2011 by UTMOST TECH LLC.

Written by Marcin Michel & Andrew J. Robinson
Edited by Lisa Day Burbaugh
Illustrations by Erin Kathleen Weston

All rights reserved. No part of this book may be reproduced or transmitted in any form or by any means, electronic or mechanical, including photocopying, recording, or by any information storage and retrieval system, without permission in writing from the publisher.

Published by UTMOST TECH LLC. www.utmosttech.com

ISBN-13: 978-1466460416
ISBN-10: 1466460415

I0429569

2012 Survival Guide

Many cultures and religions have an "End of the World" scenario, some believe it; some don't. However, many of those beliefs are rooted in one similarity: that 2012 could truly be the end of civilization as we know it! The fact that so many ancient people (notoriously Mayans and Christians) could agree on one specific time warrants our attention.

With that in mind, we are pleased to present a detailed outline on how to survive any cataclysm that might befall our world. Some of these might seem farfetched: polar reversal, asteroid impact, Nuclear Winter; while others: a pandemic, earthquakes, tsunamis, and Global Climate Change - are things we hear about on almost a daily basis. We all know the news can't predict the future.

If Ebola suddenly swept your community, would you know what to do?

If you and your family were on vacation, and an earthquake ripped through the area, how would you survive?

When Hurricane Katrina struck New Orleans, some of the people were stranded for days. Would you know what to do in such a situation? If a calamity such as that struck on a global scale, there would be no National Guard or military (or FEMA) to come to the rescue; you'd be on your own!

This book will give you step by step instructions on how to prepare for a myriad of potential disasters; how to survive them, and then how to start to re-building process.

This is knowledge you can not afford to be without!

This Survival Guide is dedicated to those individuals that take the risk of natural disasters seriously. This complete survival guide has been thoroughly researched and compiled as a vertical slice of hundreds of survival techniques, to present only practical and useful information.

Table of Contents

INTRODUCTION/GLOBAL CATACLYSMIC SCENARIOS

The purpose of this book is to provide you with the selected, logical, and "must-have" information as well as necessary resources to help you prepare for, and survive a long-term global disaster. The survival techniques and guidelines described in this book are based on several disaster scenarios grouped into the following categories:

- Earth's core and geostructural changes, natural disasters
- Solar and orbit influences on Earth's climate and weather
- Human consciousness, the evolution of warfare and conflicts around the world
- Diseases and pandemics

Each scenario ends with complete or partial extinction of human civilization, so the survival of an individual, family or group of people is not targeting temporary survival methods until rescue, but complex survival processes that are split into 3 major stages:

- Survive the disaster and evacuation (3-7 day plans)
- Settle down and secure life threatening factors (7 day plans)
- Survive long term, and re-building your community

This Survival Guide has been designed to prepare you for surviving the speculated December 2012 or similar cataclysmic events that could be significantly more devastating than other manmade or natural disasters that have happened in the past. All survivors must be more proactive during and after global cataclysmic events and must undertake more radical strategies for long term/permanent survival. All public infrastructure and utilities will be nonexistent, the atmosphere might be polluted for years, and clean water and food will be scarce. Our governments, economies, banks, and healthcare will be nonexistent as well; so nobody will be able to assist us. The human race, as well as all other living forms of life may be near extinction. We will be on our own, starting over, just like our ancestors, hundreds (or even thousands) of years ago.

GLOBAL DISASTER SCENARIOS

There are several possible cataclysmic events that are either speculated to occur sometime during 2012 or are considered by scientists as possible in general, not connected to a specific date. In any global disaster scenario outlined in this book, the survival conditions will be more severe than any previously experienced by humans. However, it is logical to assume that at least some negligible chance for survival will exist, for those who are prepared! That is what this book will help you in preparing for.

GLOBAL FLOODING

Depending on the event, a global flood can occur rapidly or at a slow progressive rate. However, we must assume that a cataclysmic flood will progress much faster than the gradually increasing sea level due to global climate change. It is unlikely that there will be enough time to build a floating home or shelter like some sort of modern day Noah's ark. Any information about such an upcoming disaster, confirmed and verified by a public announcement from the government, will most likely be available only several months ahead of time. This will be especially true in the event of a rapid Tsunami-like flood, caused (for example) by the impact of a celestial object plunging to Earth or a massive undersea earthquake. In the case of the latter, warning time could be as little as a few hours!

It will also be very difficult to estimate what areas of the planet will be impacted by the flooding (if not entire continents), until shortly before the impact of the object, or hours after the quake. What is certain is that about 60% of the world's population live in terrain that will be underwater if as little as 20 feet of water inundates those areas.

Gradually progressing flooding gives us a little more time to get organized, but at the same time it also increases the panic in the local populace. This will lead to abnormal economic activity, which will significantly reduce our chances of collecting the resources needed to build or buy any kind of floating safe house, or at least a boat that can handle any kind of extreme sea conditions. Plus, it requires some serious engineering and professional facility to build an "ark"!

That's why evacuation to a mountain area is the most reasonable decision. In the event of a short term flooding event, simply having some supplies in your vehicle and evacuating to the mountains is enough. Once the seas subside, some sort of relocation center will be established by the government - as going home will be impossible, at least in the short term. From there you can plan for the future: going home or relocating. That decision will depend on the extent of the damage to your community.

Now, the preparedness for evacuation due to flooding on a long term scale is quite another matter. Today, the oceans and seas cover 70% of the Earth's surface. Survival on the planet when it's covered by much more; for example, in the 90% to 95% range, will require quite the creative approach and good planning. When every piece of land, except 5% or 10% of the highest mountain areas are submerged, the survivors will be relying mostly on the food resources that can be found in the ocean.

So, what are the speculated causes of the Great Flood? There are quite a few scenarios, but the most speculated is a Planet Nibiru. This is a planet that was theorized to exist in the outer reaches of our solar system. It's also known as Planet X, and if Nibiru passes between the Earth and the Sun, its gravitational pull could interfere with the Earth's orbit, pulling it closer to the sun. This would result in a rapid increase in air and sea temperatures. The warmer conditions would mean massive melting of the glaciers, and much faster evaporation of water into the atmosphere. Also, a near miss of a passing planet would increase Earth's rotational speed, causing high altitude Jet Stream winds that, combined with a lot of moisture, would trigger massive super storms and great floods. When you consider the sizes of hurricanes like Andrew and Katrina, and the massive amounts of damage they caused, any storm larger than those storms would obliterate any area they impacted!

In addition, even if you ignore Planet Nibiru, scientists know for a fact that there are large numbers of asteroids and small comets scattered throughout the solar system. If one of those were to impact Earth in an ocean or sea, the result would be a massive tsunami. Depending on the size of the object, it might be too small to be detected by satellite or telescope, and thus there would be little or no warning. The tsunami that hit Indonesia in December of 2005 killed more than one hundred thousand people!

So, when the global flooding is confirmed, it would be crucial to evacuate as soon as possible, and to stay mobile, so high mountain areas can be easier to access. The panic caused by the simultaneous evacuation of several billion people world-wide all heading to the mountains is unimaginable, and for such a massive scale even a 1 year evacuation time-scale is equally unimaginable.

A Great Flood will most likely be accompanied by extreme hurricanes and huge rainfalls, followed by landslides in the highland areas. Stay in the mountain areas where the slopes are primarily made from solid rock, and build your shelter above the valleys.

Conclusion Diagram:
GLOBAL FLOODING --> MOVE TO MOUNTAINS QUICKLY
- Pack & Go - Try to collect as much cash and commodities as possible.

Plan your trip (see EVACUATION Chapter for hints) and move. Remember that in a true global crisis, money may become worthless. So, bring small, light-weight items that you can use for bartering. Something as simple as a manual can opener can be a Godsend to another person - so toss a few in your bag.

If you live on an island - move to the continent. Remember - high mountains will look like little islands after the flood. Review the mountain maps and select an area easily accessible by a road that is close to the highest peak in the mountain range.

Collect supplies, build a shelter. Remember that clean water is one of the most critical supplies. You can survive several days (even upwards of two weeks) without food, but water is vital. To that end, bring as much as you can carry, but realize that water is heavy; so look to bringing either a portable water purifier (some are even solar powered) or some water purification tablets - many are available in a pharmacy. If you or any member of your family have any sort of medical condition, be sure to bring as much of your medication as possible.

EARTHQUAKE & VOLCANIC ERUPTION

Geologists agree that major displacements of multiple tectonics plates around the globe are overdue, and it's theoretically possible that we will experience geological mayhem in the form of a chain reaction of events that the history of mankind has never experienced before. It is a commonly recognized fact among scientists that all earthquakes and volcanic eruptions documented so far might be just little "farts" of Mother Nature, and the power sleeping underground is so huge that we don't have a clear idea as to how to measure the impact of a major tectonic event.

Earthquakes are measured using what's known as the Richter Scale; it is a logarithmic scale - meaning that each number of the scale is ten times bigger than the one before. A minor quake is in the 2 to 4 range; once you get up to the 7 range, they get quite massive, and 8 to 9 is so huge as to literally move continents! The recent quake in Japan was around 9, and the entire country was moved several inches to the east. In addition, as the epicenter of the quake was in the ocean, a massive tsunami followed, leading to still more death and destruction.

In some areas - such as sections of California that lie along the San Andreas Fault - the question is not *if* a big earthquake is going to hit them, it's *when*. When one of the massive quakes hit, the last place you want to be is in a large city, and you definitely do not want to be in a high-rise building. Even in an area like Los Angeles, which is prepared for a large scale earthquake and has adequate building codes, a powerful quake will result in massive damage, a breakdown of basic infrastructure, and widespread lawlessness. This is not remotely a situation you want to find yourself in, and you definitely do not want young children exposed to it.

A possible single eruption of Yellowstone Park's sleeping giant volcano is estimated to cover the whole world under a cloud of ash for years, causing extreme winter conditions and several climate abnormalities. If the eruption took place underwater, an additional aspect would be giant tsunamis racing out from the epicenter like the ripples caused by a pebble being dropped into a pond. However, in this case, these "ripples" would be tens (perhaps hundreds) of feet in height, and would rip across the sea, blotting out whole islands and nations as they raced around the globe.

The full power unleashed from the mighty San Andreas Fault can easily destroy the US West Coast, but if the epicenter is located under the ocean the following tsunami would also cause serious damage across the Pacific Basin. In general, the risk of major volcanic and tectonic activities is a known issue in hundreds of areas on Earth, and the worst news is that we will not see it coming.

So, earthquake and volcano disasters happen suddenly, with no warning or upon very short notice. Obviously, the worst places to be during such events are the cities, especially in or close to tall buildings, skyscrapers, and areas of dense traffic where evacuation routes get jammed quickly. When giant earthquakes or volcanoes strike, there is no time for shopping for the survival gear and supplies; so it's crucial to have everything handy and ready at home. Millions of people will lose everything within seconds, they will be desperate to survive, and they will be on the run to higher ground to get as far as possible from the tsunami danger zones.

IMPACT FROM SPACE

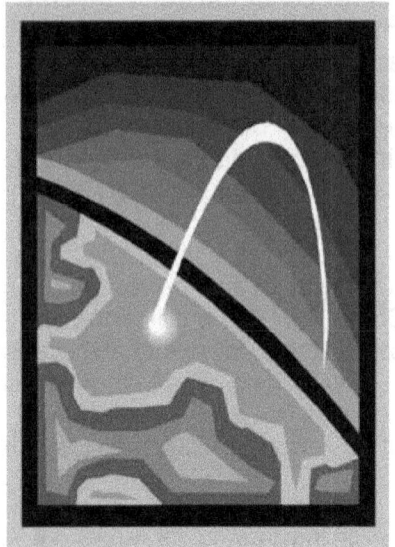

The impact of a large celestial object is probably the worst cataclysm of all; the worst case scenarios that can happen to our planet. It could be Planet Nibiru, a comet, or a sizeable asteroid. All of these are known as Near Earth Objects (NEO), and they can be anything from *merely* deadly to downright apocalyptic! We know it happened before; 65 million years ago the impact of one such NEO caused the extinction of all of the dinosaurs on the planet. As previously mentioned, even a tiny object can result in massive tsunamis and/or damage.

The chain of events after the impact is nearly unimaginable, starting from a huge explosion that will vaporize everything within a radius of several thousand miles (seconds after impact), through enormous tsunamis around the globe (hours after impact), lava storms (hours after), rapidly expanding thousand degree land fires (days after), changes in the atmosphere's composition (weeks after) to nuclear winter and acid rainfall (roughly month after).

Any chance of survival depends on the diameter of the asteroid bomb, and in general this survival guide might be simply obsolete if the diameter of the object is larger than just **5** miles. The kinetic energy of a large rock smashing into the Earth's surface at a high speed (like 20,000 miles per hour) is way more powerful than all the nuclear bombs every nation can possibly build and detonate at once.

In Arizona, there's Meteor Crater National Natural Landmark; it was caused by a rather small NEO - a nickel-iron meteorite about 54 yards (50 meters) across hitting the area about 50,000 years ago. That small object caused a crater almost three-quarters of a mile across, and its impact power has been estimated to be about ten megatons! Imagine the effect a NEO only a few times bigger would have today.

So, with that in mind, let's consider a NEO sufficiently small that survival is possible. Here are the key factors to consider: the size of the NEO and its projected point of impact. The location of the impact will tell you how close it's going to be to your community, and whether or not tsunamis will be a factor. The size will tell you a number of factors:

- The size of the firestorm at the point of impact
- The diameter of the impact crater and area of destruction
- The size of the dust cloud thrown up after impact
- The earthquakes that will follow the impact

Now, the good thing about all of these elements is that scientific experts will calculate all of them before impact. The more lead time they have, the better. They will also be able to calculate roughly how long the cloud will last, and its effects on global weather patterns. Once this information is made public, you'll need to see if your area is in the vicinity of the initial impact. If so - move - fast! Even if the impact is on the other side of the world from you, the quakes and weather changes could affect you.

To survive, you will have to move as far away from ground zero as possible. In the event of a truly huge NEO, you may have to literally dig yourself deeply underground, and stay there for a period of time! The caverns high in some mountains are the ultimate shelters. Again, this is something the experts will determine. If the impact is going to cause global tsunamis and/ or a nuclear-type winter that's going to last for years, it might be that the only chance for survival is to settle down with life support systems underground, and stay there until the "All Clear" is given.

The bad news is that there is not enough room in all the mountain caves on Earth to host 8 billion people. So, planning ahead and securing a location as soon as possible will be critical.

SOLAR FLARES & RADIATION

A Solar Flare is a natural blast of energy radiating from the Sun into space when extra intense reactions occur on the Sun's surface. This is a normal process, and the Earth's magnetic field and atmosphere safely deflect most of the Solar Flare blasts, absorbing the rest. Only trace particles arrive on the surface, and they have no real impact. However, when the giant doomsday Solar Flare hits, the Earth's natural shield won't stop it, and the planet will burn with an enormous radioactive blast.

We should have enough time to react and find a bunker to hide in, and it doesn't sound too scary to survive short periods of Solar Flare blast. The real danger and survival comes after. Due to radiation, the majority of livestock will vanish in the thousand-mile radi ground zero, and the radioactive particles traveling with the winds around the globe will cause contamination of livestock, water, soil, and radiation sickness.

The good news is that a shelter capable of protecting you from solar radiation is relatively easy to build; so survival isn't difficult. Once the immediate danger is over, the key to survival will be evacuating from the irradiated zone. When the reactor in Chernobyl exploded, the ground around it for miles was contaminated, and will remain so for decades. That was a small blast from a manmade reactor; imagine the area affected by a single huge Solar Flare - it could encompass whole nations or even entire continents!

Another possible event that could blast the Earth with solar radiation is a reversal of the poles. This is where the Earth's magnetic field switches ends. This may sound fantastic and out of some sort of science fiction story, but it actually takes place about every 400,000 years. Scientists are still unsure as to *how* it happens, but they know (via fossil records) that is *does* happen.

Scientists are also unsure as to what the affect this pole reversal will have on the planet. In a worst case scenario, the Earth might start spinning in the opposite direction, which would cause massive tornado-type storms, electrical storms, earthquakes, and widespread tsunamis. To survive such events, an underground bunker in the solid mountains of Asia and the Appalachians would be the best bet.

NUCLEAR WINTER & DISEASES

In the case of global nuclear winter or global diseases, the highest chance for survival is to move as far as possible from the initial blast zones or initial outbreaks. The uncontrolled, rapidly expanding diseases usually spread across the continents within 2-3 weeks, and then migrate with traveling people or animals, reaching global level in several weeks. To a large degree, the rate of spread is dictated by the disease's incubation period. The longer that timeframe the further people and/or animals can travel before symptoms appear and the subject is either quarantined and/or so sick that it can't continue to travel.

Nuclear winter on a global scale would most likely cover the entire planet. However, in both cases there might be remote areas where the radiation/contamination and/or disease are non-existent or negligible. Massive atomic explosions at ground level (following the nuclear exchange associated with a war) will cause an incredible amount of dirt and debris to be thrown up into the atmosphere. Millions of tons of radioactive dust swirling around will block the sunlight from a few months to several years!

As a result, the planet would get extremely cold; vegetation would wither and die (being unable to grow due to a lack of photosynthesis). After that, animals would start to die off - first the herbivores (plant eaters), and then the meat eaters who rely on the other animals for food. Vast areas would end up ice cold deserts or resembling the "Dust Bowl" of the 1930's. This was a region of the United States where - due to drought - the plants died. Without plants to hold down the soil, the topsoil literally blew away, and the area was reduced to a giant desert-like area.

In addition, the radioactive dust would settle on the ground, in the water, and the lighter particles would linger in the air.

This would lead to people breathing them in, drinking them, and even walking through them. As plants and animals (in the areas where the sunlight returned) began to return, they'd incorporate the radioactive material into the food chain. People would be faced with the prospect of either eating contaminated meat and vegetables or going hungry.

As a side note, following the Chernobyl disaster, the Soviet Union made the decision to distribute the contaminated food from the regions affected by the fallout. It was reasoned that so much food should not go to waste, but to minimize exposure they sent the food all over the country. The thinking was - a little radiation wouldn't hurt people. On the other hand, the people in the immediate area of the reactor had fresh, clean food shipped in (especially for the children). Also, children were kept indoors as much as possible to minimize exposure to the contaminated air and soil. The government even built a re-creation of a normal healthy forest *inside*, so the children could see what the world *should* look like!

That was *one* small disaster in *one* country - and a country big enough and powerful enough to absorb it. Imagine such a disaster on a global scale! What would happen to the contaminated food? Would governments take such care to insure people got as little exposure as possible? Would there be adequate safe zones for children to live and grow up healthy? These are things that *you* are going to have to take care of; that *you* are going to have to think about and plan for.

Now, you might say, "But wait, the Cold War is over; the US and the Soviet Union aren't pointing missiles at each other anymore." In fact, there isn't even a Soviet Union anymore! That's true, but there *are* still many nations that have atomic weapons, and there are also terrorists determined to get their hands on a "Nuke" to use as a weapon. Both Pakistan and India have "The Bomb", and they are forever squabbling over something. Is it such a stretch of the imagination to picture a scenario where a terrorist group sets off a bomb in Mumbai or Islamabad?

What would be the result?
The two nations could easily go to war, shoot off their nukes at each other, and plunge the world into a nuclear winter. Even a limited nuclear exchange would throw up tons of debris that

could easily lead to a nuclear winter of one to three years. The safest places to be are the Southern Hemisphere and the far north; both are places out of the main air currents, and thus will get minimal fallout.

Super Virus or Bacteria
Despite all the wars that people have fought over the millennia, the greatest killers in all of history have been the diseases cooked up by Mother Nature. The worst of these has been the Black Plague (also known as the Black Death) of the 14th Century. This was the bubonic plague, an incurable bacteria spread by the fleas living on rats. It spread across Europe, and in some communities fully 90% of the population was wiped out! Over the course of the 100 years of the Black Plague, about 30% of the entire population of Europe died. While the bubonic plague is no longer a threat, as recently as 1919 (following World War I), the Influenza Pandemic spread around the world, and killed tens of millions of people.

While influenza is also now under control, there are new diseases appearing all the time. We've handled Bird Flu and Swine Flu, but those are nothing alongside diseases like Ebola and AIDS. In the case of Ebola, it is a swift and deadly disease. About the only good thing that can be said about it is that the symptoms manifest within a few weeks, which means the area of an outbreak can be quickly determined - and isolated. With AIDS, it takes longer, but remains incurable.

Beyond the natural diseases out there, there are also the artificial ones. Various governments around the world are experimenting with different types of biological organisms. In some cases, these are weapons; in others, they're treatments for diseases. In addition, companies are also splicing genes and altering cells on an almost daily basis. They're trying to create cures for diseases, organisms to create insulin, crude oil, and a host of other useful products. However, if just one of those new biological entities should get out of its lab and into the environment - who knows what might result? A deathly plague *ten* times worse than anything previously seen could be unleashed on the world!

We have to wonder, will 2012 be the year the next Black Plague emerges? As international air travel becomes easier and easier, people can get from an infected area to virtually anywhere on Earth in a few hours. This just about guarantees that they can become infected and travel anywhere else before showing any symptoms of the new disease. As a result, a disease can spread almost instantly across the globe, especially to large, overcrowded cities teeming with people.

That final point is the main clue to surviving this scenario. In the event of any sort of outbreak, you need to follow these critical points:

1. Keep all members of your family as fit and healthy as possible at all times. You never know when a new disease is going to appear. To that end, here are two things to consider: First, avoid taking too many medications, as this can lead to you building up a tolerance to those meds, and then (when you need them to work) they become less effective. Second, when a doctor gives you medicine to take, be sure to take it all - use it up! If you're told to take a med for 30 days - take it for that long. Many people stop taking medicine once they feel better. However, this does not mean the disease is gone from their body, and this has led to the new super strains of diseases we've been hearing about. The diseases build up a tolerance to the medicines, and next thing you know we've got a tougher "bug" to kill. Don't contribute to the problem!

2. Try to live in as remote, rural an area as you can. Now, this may be difficult owing to the type of work you do and/or your family situation. So, in that case, have an evacuation plan in place so you can relocate as soon as an outbreak is confirmed. Maybe you have relatives and/ or friends who live in a remote area. If so, make plans in advance concerning you coming to live with or maybe just visit for an extended time.

3. If all else fails, have some camping equipment ready and simply take your family into a rural area and pitch a tent or set up a camper.

4. Make sure the area is clean, and has an adequate supply of fresh water - or that you bring some type of water purification system.

5. If the area you're going to relocate to is in a cold climate (or has severe winters), make sure the home has adequate heat. Even an old-fashion wood burning stove can be enough - just ensure you have enough firewood.

Dealing with hot summers can be as easy as dressing in light clothes, staying out of the sun, and drinking plenty of fluids. A good air conditioner is best, but in a worst case scenario, you may have to make do with simple fans.

6. Keep your contact with the outside world to an absolute minimum. Your immediate neighbors should be okay, but anyone from a major city should be avoided.

7. Both before an outbreak and once you relocate, try to eat only fresh healthy foods. Remember, many so-called "Factory Farms" give animals antibiotics as part of their daily food! This helps the animals to fatten up, but it also makes any bacteria and/or viruses in their meat more drug resistant. If you eat too much such meats and then get sick, normal medications won't work for getting you well.

A key element to remember in any sort of pandemic is that the cities will soon be choked with hundreds (if not thousands) of dead bodies. Now, think about what that's going to mean. Within a matter of days - especially if it's during the summer months - the bodies will start to decay and stink. This will lead to still more diseases spreading. Therefore, leave the cities as soon as possible.

INFORMATION & SITUATION ASSESSMENT

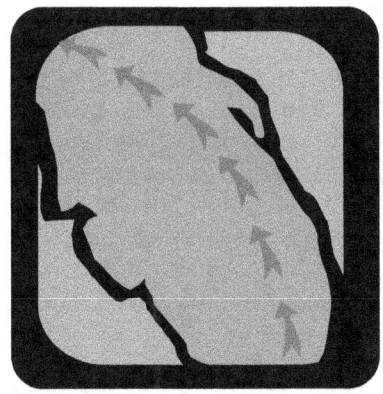

Here is the single most important cause of failure when a disaster strikes: Panic! This is something you always need to remember. In a crisis - stay calm. Think about any disaster movie you've ever seen. Now, granted, a movie does tend to be overly dramatic, but the underlying concept is correct: those who keep their wits about them, who stay calm and think, those are the people who survive.

When you find yourself in a difficult situation, here are the three things to consider before you make a decision:

STOP -- THINK -- ACT

Depending on the scenario you find yourself in, you may have anywhere from several days to a couple hours, or only minutes to get ready. So, it is crucial that you have an evacuation plan that's adjustable, and that you can modify the supplies you plan to take. Leaving with five minutes notice means grabbing just the bare essentials and going. If you have a couple days, you can tailor your supplies to the situation and where you're going. Taking provisions to an underground shelter or fortifying your house are quite different from heading up into the mountains to wait for a tsunami to pass.

INFORMATION
These days, information is one thing we always have plenty of. Between the networks, cable, the Internet, and even social networks like Twitter, you can always get information on absolutely anything! Unfortunately, that's also the problem; where there's lots of information, there's also lots of mis-information! Just look at recent events: the bin Laden operation handled by the US Navy SEALS. Initial reports told of a firefight; then it changed, and now we know it was quite

different. For that matter, look back in history to see examples of information gone wrong. Orson Wells did his famous Halloween radio broadcast of "The War of the Worlds", and thousands of people thought it was real!

In the case of a disaster, depending on the size, scope and timeframe of the event, information could be sketchy - at best. So, before you decide on how to react, verify the information about the event as best you can. If possible, compare reports being given by multiple information sources, and look at where they agree and disagree. If one source says to use Main Street to leave the city, but three other sources (including your brother on his cellphone) tell you that Main is a mass of confusion - find an alternate route.

At all times stay informed about changing traffic patterns and available evacuation routes. In an emergency, a single stalled car can lead to total gridlock. Local knowledge of the secondary roads and actual observations of the changing disaster driving situation are key for making the right decision on evacuation routes. Have a GPS in your car, maps (detailed) of the area, and really know your community. Very often, people assume that the major highways are the best way to go. In a disaster, the government may even close the lanes of a highway that come into a community so that people can use they for evacuation. However, in a true panic situation, even those additional lanes may not be enough. This is where knowing the access roads and little-known trails can be critical in getting away quickly.

Making a proper assessment of the situation will greatly increase your chance of survival. Once information on the large scale disaster is made public it will most likely ignite a panic. So it's extremely important to stick to logical decision-making, and do not blindly follow the crowd.
For example, when there is information about a sudden, unexpected disaster that requires immediate evacuation, you should quickly take the minimum immediately available survival items and supplies, and evacuate as soon as possible. There will not be time to go shopping or pack tons of supplies for long term survival.

COMMUNICATION

Your ability to contact and communicate with other survivors will be vital to your existence, and perhaps your overall sanity. Lots of people like to be alone for a while, but long term isolation is detrimental to mental health. So, having some means of communicating with other people (cellphone, computer, ham radio etc) is vital.

In addition, having the ability to stay informed and get updates about the continuing unfolding events of the disaster will help you better cope with ongoing dangers.

There is a high risk that none of the telecommunication infrastructure will be in existence following the disaster. In the event of a nuclear blast, the EMP (Electro-magnetic Pulse) will knockout much of the electronic devices within the blast area. The dynamo powered weather and emergency radios (UKF) that require basic, easy to rebuild infrastructure, and CB radios for short range communication, will be very useful, especially during the evacuation process.

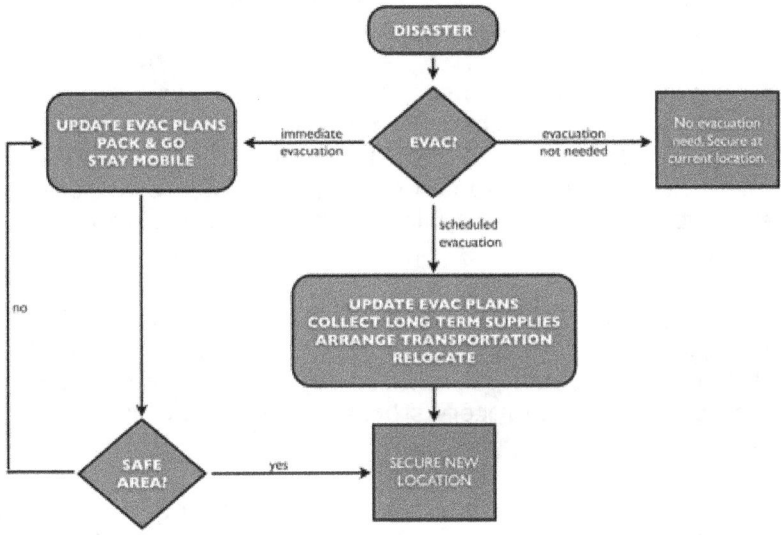

Figure 1. Evacuation Plan Diagram.

URBAN SURVIVAL - PREPAREDNESS

Now, not everyone can afford the luxury of building an underground bunker specifically designed and equipped to survive a coming disaster. You might also already be established in a city - your life, work and family are there. Also, there's the issue of time - time spent out in the wilderness hold up in your bunker. As we don't know when one of these disasters is going to hit, you can't very well go up into the mountains or move into an underground bunker and stay there. Remember, when the disaster hits, you may need that outpost of yours to live in for months or even years. What good will the bunker do if you live in it and use up the supplies? That place is for survival during and after the event.

So, if you can't afford to set up such a place, or moving out of the city just isn't feasible (for any of the reasons listed above), you may face the prospect of staying in the city. Urban survival refers to one's ability to continue living in their own home for extended periods of time without the comforts of modern day conveniences.

To do so, you'll need to prepare yourself and your family to do what is known as "living off of the grid". This means having no electricity, no running water, no telephones, no police and fire protection, and no grocery stores, hardware stores, and so on for those needed supplies.

Advanced preparation is the key to your continued existence! If you can't evacuate, you *must* survive, and here are the critical factors to remember:

If the disaster is something that is going to cause a long term disruption of government, basic utilities and infrastructure, and general chaos, then you're going to have enough time to learn and educate yourself about producing food and water, as long as you have enough short term supplies to survive the first several weeks.

With non-existent infrastructure, the urban environment can still be the best place to survive, as long as the strategies of basic survival and personal protection are applied to everyday living.

Destroyed cities will hide many treasures, and we don't mean gold and silver - these will be worthless. No, it'll be such things as canned foods, fuel, medicines, building supplies, and so on. However, at the same time, the level of destruction in an urban environment might be way more dangerous than the countryside. This will be due to the risk of collapsing buildings, bio-hazards, gas leaks, general lawlessness etc.

To prepare for urban survival, stick to the following core rules:

1. Enhance security of your living space. Turn your house into a bunker, as the crime rate and the search for food among other survivors may become one of the most dangerous post-disaster threats. Install dead bolts, grates or bars on the windows, and other heavy duty protection throughout your home.
2. Collect enough long term food and water to survive in-house for at least 60 days. Some disasters will force you to wait in the sealed homemade bunker for a long period of time. Currently available MRE (Meals Ready to Eat) or canned food can be good for as long as 7 *years*. Purchase additional water filters, such as reverse osmosis technology that will allow you to purify water once your initial supply runs out.
3. Prepare enough plywood and non-respiring foam or foil, and cover and duct tape completely all doors, windows, and airways to seal your house in case of heavy air pollution, chemical or biological hazards.
4. Equip yourself with survival kits, medicines, tools, personal protection, and nutritional supplements. And yes, "personal protection" can mean a firearm! As much as you might hate the thought, you may have to defend yourself and your family from marauding gangs of looters and/or criminals.

5. Extra batteries, a power generator, and a supply of fuel might be very helpful, but in general you shouldn't rely on any electric devices other than the ones needed for life support (e.g. the medical condition of a family member).
6. Prepare evacuation plans, and select a mobile set of the most needed survival items and supplies. Temporary evacuation might be needed when the disaster strikes, if possible. After which you might be forced to stay on the move for some time, until the situation clears, and you can get back to your urban shelter.
7. Without any sort of social structure, the mentally ill, criminals, and all manner of sociopaths will be on the loose. Even among so-called "reasonable people", they will be desperate for food, water, drugs etc. They may organize themselves into gangs and start looting every building they find. To help "camouflage" your home, make it look like it's already been searched. Throw trash, clothes, pots and pans, and all manner of household items on the lawn. If you're lucky, the gangs will pass your place by.

Any global disaster may turn all of the cities (especially those situated close to the shoreline) into inhabitable graveyards; so urban survival might not be available. Most likely, survivors will be looking for shelter as far as possible from the disaster epicenter. This is where having a home well-prepared will come in handy; you can stay put while others flee the community.

A critical factor to consider is the mental health of you and your family. Despite all the preparation you might go through, when the actual disaster comes, you have no way of knowing how you and your loved ones will react. Some people can fall into a deep depression when they realize that their world has essentially ended. All that they knew, in terms of their modern life, disappearing overnight can result in people becoming what is known as dissociative: they enter a trance-like state where they are unable to function. If that happens, you must be prepared to carry that person (or persons), or leave them behind!

In the long term, you may have to cope with family members being bitter, angry, depressed etc. To go from a high tech society with every modern convenience to one where they have to work hard all day - *every* day - just to have the basics of survival can be very tough for some people - particularly young people who have never known want or hardship. This is why some practice "drills" can be a help - go camping with your family and (as they say) "rough it" for a while. The more preparation you can do before disaster strikes, the better your chances for long term survival.

Finally, you'll have to learn to be wary of others - even people who seem completely normal. Some people will be dealing with great depression at the loss of their loved ones and way of life - this can lead them to be suicidal, and potentially dangerous if they decide to "free" you of the "pain" of having survived the disaster. On the other extreme will be people who see their survival as a sign that they are special, important, selected (anointed) by God. Such people can be quite dangerous, as they may see themselves at the rightful leaders of the "New World Order", and may become violent if they believe you are defying their "divine power".

THE PERFECT SAFE HOUSE

The perfect Safe House is dedicated to zero hour survival and long term survival after the global disaster strikes, and it must be designed with the following features:

- Located as far as possible from ground zero, and in general far from the shoreline, at least 1500 feet above current sea level, located in a rural area far from any urban infrastructure, ideally in a high mountain area.
- Located off the road, possibly at least 2 days' hiking distance from the nearest neighbors and civilization, as well as invisible from the beaten track.
- Located above the road level, far from any potential mud slides, ideally attached to stable rock or built inside a cave.
- Located within daily walking distance from a creek or well (access to clean drinking water), and the forest (access to firewood and wildlife). If possible, have some fruit trees, berry bushes, and other natural food sources in the area - either those growing naturally, or (if time permits), plant some.
- If you're very lucky, find a place with an artesian well; this is a well under natural pressure, and thus you can attach a pipe to it and have running water. Granted, the water pressure may be very low, but it's better than nothing.
- Set up a beehive. The bees will help to pollinate your garden, and you can collect honey from the hive.
- The house must be single level, possibly with a basement, or safe room drilled in solid rock underneath the house.
- It must be constructed on solid rock, with strong wood or steel beams, solid steel roof, and heavy duty shutters on all windows.
- The house must be easy to seal, so the less windows and doors the better.
- Possibility of installing solar panels and/or a wind turbine (prepared installation kit).
- Possibility to block access to the house, for instance by pushing rocks on the road leading up to the house.
- Possibility to escape through a concealed "back door" route to the nearest emergency refuge.

Of course, buying a place like that takes a lot of money, but fortunately the best places to survive 2012 *aren't* the ones with the high price tags now. A place with a great sea view, ready access to nice restaurants and trendy shops, and near a large urban center will count for nothing when the tsunamis, hurricanes, firestorms, and so on start rolling in.

However, when information about the global disaster becomes public, verified, and a real threat, you won't have to worry any longer about money and legal procedures to purchase land for your safe house, as the whole economy will be frozen.

The essential features of the geographical location of the Safe House will be strictly dependent on the type of disaster and the ground zero location. In general, most of the disasters that can happen with a global impact will require us to find a new home at least 1000 feet above sea level, 100 or more miles from the coastline, and well hidden away from major evacuation routes and cities.

It's almost certain that survivors will eventually became dangerous, ready to kill other people just to steal from them whatever might be helpful in their desperate fight for survival. When all the life-threatening factors are under your control, and you can see the light at the end of the tunnel in your long term survival, the human factor might be the most dangerous one left. That's why it's important to find a location where your Safe House will not be visible and easily accessible by other humans, and far enough from the nearest major town - at least a 2 days' hike.

Now, you might say, "But wait, I'm not rich; I can't afford to buy land and/or build a safe house! How can I protect myself and my loved ones on a small budget?"

In that case, go with a very minimal survival plan: camping. Have tents, sleeping bags, and all manner of survival gear prepared and ready. If you can't afford to even buy a piece of land in a remote area, just find two or three good sites that meet all of the criteria outlined in this book. Know how to get to each of them, and then - when disaster strikes - head to the one best suited to your needs. In the event roads are clogged with other evacuees - or someone else has taken your primary safe location - go to one of the alternates. This is why you want a couple lined up.

Don't worry about squatting on someone else's land (or maybe public lands in a park); in the event of a global disaster, private property rights are not going to count for much. So long as you're able to defend yourself and your family, that's all that's important.

Granted, merely camping out in the open is not the best means of protecting yourself and your loved ones from the disaster, but at least it's something. Also remember this: do not sleep on the bare ground; you need to be up off of it on some type of padding. It can be something as simple as a sleeping bag or air mattress, but it needs to be something. The ground tends to cool off during the night, and you will lose valuable body heat sleeping on it.

EVACUATION - PREPAREDNESS

THERE ARE THREE THREES TO REMEMBER:
- 3 Weeks without Food means death
- 3 Days without Water means death
- 3 Minutes without Oxygen means death

When the information made public about an upcoming disaster has been assessed and verified, you will have to choose your evacuation plan accordingly. Choose is a key word here; as there might not be enough time to prepare the evacuation plan when the disaster actually strikes. It's important to plan ahead, and get familiar with key aspects of each evacuation scenario. Keep your evacuation bags with survival kits, tools, and basic supplies at home, or wherever you spend the most of your time. If possible, keep some items in your vehicle - such as tools, tents, and other non-perishable gear.

There's going to be great uncertainty regarding your ability to obtain gasoline along the way. What stations are open will be swamped with people trying to fill their tanks. So, don't just keep your tank full, also have a couple gas cans in your car, and a hammer and screwdriver. In the event of a true global disaster, you can scrounge gas as you go. Simply stop at any abandoned vehicles you find along the way, use the hammer to smack a hole in their tank with the screwdriver, and sit a gas can under the hole. It isn't a perfect means of draining a tank, but it will work. Do this to every vehicle you find and you'll be able to keep yourself well-supplied with gas. Also, be sure your gas cans have a pour spout and you have a funnel. You'll need one or the other to allow you to fill your tank safely - and not lose any of that valuable fuel, and the funnel will help you collect the gas as it flows from the other vehicle.

In terms of a vehicle, some type of camper is the best. While it won't get as good gas mileage as a compact or hybrid car, it has several pluses. A camper will have more space for storage, you can have food and refrigeration, you can cook, and it has beds. You can also easily secure bikes, motorcycles, and equipment to the outside, and the sheer bulk and size of the camper will be helpful if you have to ram through a barricade or road block!

There are two types of evacuations that you might face: immediate and scheduled. Let's look at details for both.

IMMEDIATE EVACUATION

This is a quick re-location with just enough time to pick-up pre-purchased items that are essential to survival for the first weeks after the disaster. After re-locating to the safe house area, and surviving the strike, blast, outbreak etc you will have to find a source of fresh water, start building a shelter, start hunting for food, and then (when the basic life-threatening factors are eliminated) start building a Safe House for long term survival.

Here is a breakdown of the key elements of the evacuation:

- Evaluate the emergency evacuation routes and traffic patterns between your current location, the place where you need to pick up your basic survival items, and your destination. Remember the three steps: Stop, Think, Act - and then select the fastest and safest means of transportation to your Safe House location.
- Pick up your survival kit and pre-purchased basic supplies. Take your car, but stay mobile - pack everything in light carry on bags or backpacks, as you might need to abandon your vehicle in a hurry.
- Try to avoid public transportation - anything public means it'll be jammed with panic-stricken citizens during a large scale evacuation.
- Don't waste time shopping during an immediate evacuation - it's better to get on the road, get past the congested traffic areas, and then collect remaining supplies on your way to the safe house location. Not only is it a waste of time to try and buy things, but everyone else will be trying to do this. As a result, the stores will be jammed with people, vital items will be in short supply, and prices will be much higher than normal. It is quite common for merchants to jack up

prices during a crisis.
- Stay informed about the disaster - be aware of the chain reaction of events that might be changing often, sometimes minute by minute. By keeping a radio and/ or cellphone with Internet access on and monitoring the news, you can quickly adjust your evacuation route, as needed.

SCHEDULED EVACUATION

This refers to a situation where the information on an upcoming disaster, including its impact location and type of disaster is known at least a month ahead of time. Depending on the schedule, you might be able to plan multiple transportation trips and some reasonable time to build your safe house in a new location.

Here are the critical items to consider:

- Evaluate the government issued evacuation plans, and verify those plans and schedule with your destination, transportation needs, and supply list. Choose the earliest possible evacuation date. Remember, do not worry about work, school, holiday plans etc. You are in a life or death situation; your kid's birthday or school play, or even your wedding anniversary can wait!
- Pack only the items that can be useful in a Safe House, keeping in mind the possible future non-existence of vital infrastructure. Don't take electronics, except for portable and easily re-chargeable devices or those with adapters to plug into your car's power, such as GPS, cellphone, weather radio etc. Don't bother with furniture, it's too heavy. Don't bother with fresh or frozen food, the former will quickly spoil, and the latter will thaw and spoil. Now, if you want to grab some fresh fruit to eat on the trip, that's okay; such items are high in fiber, natural sugars, and vitamins and minerals. Canned foods and long term meals are the only food stock to bring a lot of, and don't take any heavy items.
- Pack at least 7 gallons of water per person, or as much water as you can carry and a water purification system. Remember, a gallon of water weighs about 8-1/3 pounds; so 7 gallons is almost 60 pounds! For a family of 4, that's upwards of 240 pounds. Can your car handle that?

- Equip yourself with all the survival kits and tools listed in the SURVIVAL KITS & TOOLS chapter.
- Withdraw as much cash as possible from bank accounts. Collect only the immediately available funds that you'll need to spend during the evacuation. Remember - money, gold, silver (even diamonds) will be worthless compared to food, medicines, water, or even bullets during your long term survival. However, do bring your bank book, check book, passports, credit cards, money card, ID, and other important documents. In the event the crisis turns out to not be long term, you'll need those documents to get your life back in order.
- Equip yourself with self-defense and hunting weapons. If you can't get any firearms, buy a bow and arrows, tear gas (and gas masks for yourself and your family), hunting knives, and similar non-lethal protective gear. In terms of firearms, you'll want both rifles and handguns. The rifles will help in hunting and protecting your home at long range. But, a rifle is not useful up close. If - worst case scenario - your safe house is attacked by violent people, handguns are the best defense at close range.
- Load shelter / safe house building tools and basic hardware supplies, such as ropes, nails, bolts, wire, glue, duct tape, and similar light construction elements. Take only items that you will be able to carry on your back in rough mountain terrain. All heavy items: lumber, plywood, bricks and mortar, ladders etc should already be at the safe house.
- Pack a full supply of any prescription medicines that you must have, and add a supply of commonly used drugs.
- Pack first aid kits, at least 3 full kits per person.
- Pack only light and portable personal gear and only durable, warm, and comfortable cloths. A ski outfit and trekking boots will be much more useful than a coat and flip flops.
- Don't forget to take area maps, and fill up your car's gas tank, as well as fill any extra gas tanks - either mounted in the car or portable ones. Any gas left over once you arrive at the safe house can be used later.
- If you believe you'll be able to return to your home after the disaster - protect your home by installing shutters, sealing the doors and windows, and cut off the utilities (electric power, main water valve etc).

EMERGENCY POWER

A gasoline powered generator will be useful in the event of short term power outages. Depending on the fuel availability, it can be used to power low energy consumption devices, such as lamps, a microwave oven, or radios. None of these devices consume a large amount of power, unless used at the same time. In the event of a long term evacuation (or you are unsure as to how long you're going to have to stay at the safe house), you'll want to try and minimize your use of the generator to conserve fuel. To this end, consider having oil lamps and candles in the safe house, and use a wood burning stove for all cooking. The microwave oven is convenient, but - for a long term evacuation - every watt of power and every drop of gasoline will be as precious as gold! For the radio, make sure you have plenty of batteries.

SURVIVAL KITS

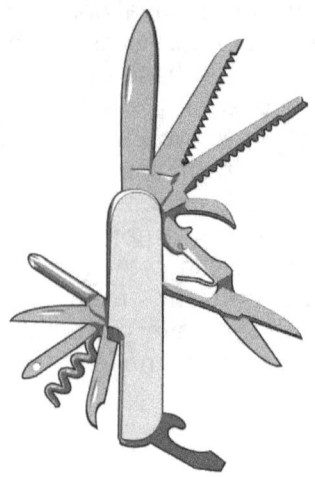

Preparing a solid and useful Survival Kit is the most important step to increase your chances for survival in case of disaster. Carefully selected items for the kit will not consume any significant space in your closet, and will definitely insure that you and your family survive.

Once disaster hits, you won't have time to shop or search for supplies. But if you've gathered the critical items in advance, your family can endure an evacuation easier.

ULTIMATE SURVIVAL KIT

--- LIST OF ITEMS & DESCRIPTION --

So, let's look at some basic items you should either stock beforehand or get just before evacuating to your safe house:

1. Generators: the size will depend on your income, the size and scope of your safe house, and the size of your family.
2. Water Filters/Purifiers: see separate section later for a full discussion of this subject. Basically, go for a system as simple as possible to minimize the need for repairs and replacement parts.
3. Portable Toilets: either a chemical one or a simple seat that you can use when setting up an outhouse. In the event of a long term evacuation, an outhouse will work best. Chemical toilets have to be emptied and re-filled, and that will become difficult over time. If you have the money and time, you can install a septic system. But, keep in mind, such a system needs running water. Also, be sure you set up a septic system and/or outhouse *downstream* of your house. The waste water

will filter into the groundwater, and you do not want to contaminate your well water.

4. Guns, Ammo, Pepper Spray, Knives, Clubs, Bats, Slingshots: all of these are weapons that can potentially save your life and the lives of your family.
5. Duct Tape: this is an all-purpose repair item that can come in handy when a pipe bursts, a tent tears etc.
6. Cook stoves (Alcohol, Propane, Kerosene): each of these has its pluses and minuses; the key drawback to each is the need for fuel. However, in the short term they can allow you to cook decent meals yet use a minimal amount of fuel.
7. Fuels: gasoline, oil, kerosene, LP gas etc. Each of these has its uses; you just need to decide which one(s) are best for your situation, and then stock up.
8. Garden Seeds: if you have to evacuate for a long period, and possibly start over, growing your own food is going to be important. Stick to regular plants, use non-hybrid plants, and those that grow naturally in the area. If you live in a northern area buying seeds that need a tropical climate to thrive would be a mistake.
9. Feminine Hygiene Products: if anyone in your group is a woman in her childbearing years, she's going to need these. In a disaster, the last thing you need is someone menstruating and slowing everyone down.
10. Skin products, hair care, deodorants: if you're going to be out in the wilderness for a long period, you need items to help your skin stay healthy, your hair stay clean, and your body not stink.
11. Toothbrush/paste, mouthwash, dental floss, nail clippers: all of these tie-in with the above item. You need to have the ability to keep your mouth as clean as possible, and your nails trimmed. A simple toothache or cracked nail can be painful and distract you from the more important issues around you.
12. Shaving supplies (razors, creams, talc, aftershave): unless you're okay with growing a beard, which may be the case if you have to stay in your safe house for a long time, bring at least some basic shaving items.
13. Waterless & Anti-bacterial soap (saves water): it is vitally important that you (and your family) keep your hands as clean as possible. You're going to be working in the wilderness doing all manner of things - digging in the garden, gathering food, fishing and hunting, and

so on. Any germs on your hands can easily get to your eyes, on your face, and in your mouth. Normally, this would necessitate a simple visit to the doctor; in time of disaster that will not be possible. So, keep your hands clean - it'll be a big step in staying healthy.

14. Baby wipes & baby oils etc.: if you have any infants in your family, you'll need a full line of baby products.
15. Diapers/formula, ointments/aspirin, etc.: don't both with disposable diapers, go with cloth. Given the uncertainty as to how long you're going to have to be away, cloth can be rinsed out and used again. If at all possible, breastfeed your baby; not only is it healthier and gives your baby much-needed natural immunizations, but it'll save on the need for formula.
16. Propane Cylinders: large ones will allow you to maintain a stove and heater for a longer time, but they are also heavy and bulky. If possible, put these in the safe house long beforehand.
17. Flashlights, light/glowsticks, torches: flashlights are key in being able to move about at night. Glowsticks are handy in a pinch - just snap and shake. Torches are good if batteries run out, and some have chemicals that will keep insects away.
18. Lantern Hangers: this may seem obvious, but there are going to be times when you need to hang a lantern up.
19. Candles: always handy for providing light.
20. Matches: make sure they are the "strike anywhere" type so you can easily light them.
21. Lamp Oils and Wicks: as with candles, these are good to have when power is in short supply.
22. Charcoal and Lighter Fluid: if your gas stove runs out, being able to cook on a grill is always good.
23. Carbon Monoxide Alarm (battery powered): if you have a gas generator or any sort of motors around, you need to be careful of carbon monoxide - it is silent and deadly. Even a slight exposure can leave you disabled and weak for weeks! In a crisis, this is not how you want to be.
24. Pump Repair Kit: if you have any sort of gas grill (Coleman etc), have this on hand in case of damage to the grill. The same goes for a water and/or fuel pump.
25. Gasoline Containers (Plastic & Metal): you have no idea how long the crisis will last; so have plenty of fuel on hand. Ideally, buy a canister a week and take it out

to your safe house. Over time, you'll build up quite a supply.

26. Hand can openers, hand eggbeaters, whisks, and other cooking utensils: to minimize your power usage, use as many hand operated items as possible. Electric can openers and mixers are convenient, but costly.

27. Cast iron cookware: this may seem strange, but cast iron is not only sturdy and efficient, but the iron is healthy.

28. Cook books: even if you know how to cook, having some of these books around is a good idea. You can start to vary your diet, and they have conversion charts for various measurements.

29. Insulated ice chests: use to transport any perishable items to the safe house.

30. Aluminum Foil: get both regular and heavy-duty foil. They're good for cooking, storage, and as a barter item.

31. Garbage Bags: you can never have too many. They're useful for everything from gathering trash, and nuts and berries to covering a small hole.

32. Toilet Paper, Tissues, Paper Towels: each is vital. If you're going to be away from civilization for any length of time, you're going to need plenty of toilet paper (unless you plan on using leaves!). Tissue paper is good for anyone with allergies, and paper towels are always good for a simple clean up.

33. Plastic plates/cups/utensils: they are light-weight, and easy to clean.

34. Paper plates/cups: buy as much as you can and stock your safe house. In the event a means of washing dishes is not available, you'll need these.

35. Clothes pins, a clothes line, hangers: unless you have a really well-stocked safe house, you are not going to have a washer and a dryer. So, you'll be washing clothes in a stream, and then hanging them up to dry. Warning: if you're going to use the stream/pond for water as well as washing, make sure you wash your clothes downstream from where you draw your drinking water.

36. Washboards, Mop Bucket with wringer: use all of these for basic cleaning and laundry.

37. Laundry Detergent: stick with a basic liquid detergent that you just mix with water.

38. Bleach: make it plain, *not* scented, and with 4% to 6%

sodium hypo chlorite. You can use it to help purify your water - if need arises.

39. Fire Extinguishers: without a fire department to call on, you're going to have to handle such things yourself. As an alternative, put Baking Soda in each room.

40. Batteries: get them in every size, and be sure they have a long expiration date. If possible, get re-chargeable batteries and an adapter to connect to any power system you may set up - a generator, solar panels etc.

41. First aid kits: there are plenty on the market that are specially designed for long term survival.

42. The Boy Scout Handbook: it's full of useful information for surviving in the wilderness.

43. Survival Guide and Emergency First Aid Books: like the Boy Scout Handbook, these can be a godsend when faced with a dangerous situation.

44. Farmers' Almanac: this is full of useful information on the tides, the seasons etc.

45. Water Containers: get plenty and in lots of sizes, but not any that are too large. Remember, water is heavy, and you're going to have to carry them. To drink out of, use only small, food grade containers. You don't want the chance of spilling even a drop of water.

46. Garbage cans: get plastic ones, and ideally ones with wheels. Not only can you haul your trash away from your living area, but they're great for storage, and even hauling water.

47. Atomizers: these are useful for cooling off in the heat. Just fill with cool water and spray on your extremities and the back of your neck.

48. Writing paper/pads/pencils, solar calculators: not only will you need to keep notes on important information (directions, critical dates, locations etc), but you may get bored as time passes. Being able to write will help to pass the time. The calculator will be useful in all types of math problems: fuel consumption, formulas for cooking, measuring out medicine etc.

49. Journals, Diaries & Scrapbooks: this ties-in with the above item. Boredom can be mind-numbing, especially as the weeks turn into months! So, having something to occupy your time is important.

50. Board Games, Cards, Dice, and Books: simple, basic recreation is vital in keeping your spirits and morale up, and keeping the children occupied. Also, a set of classic

literature will be invaluable in passing the time.

51. Reading glasses: if you or anyone in your party needs glasses or uses contacts, have a ready supply.
52. Scissors, fabrics & sewing supplies: as time passes, clothes will need mending. Also, as the children grow, you'll need to fashion new garments for them.
53. Tarps/stakes/twine/nails/rope/spikes: all of these are helpful in covering, tying down, securing, repairing etc all manner of things.
54. Sleeping Bags & blankets/pillows/mats: if you don't have time to put beds in your safe house, or you have to evacuate quickly to another location, these are going to be needed in order to sleep well.
55. Cots & Inflatable mattresses: same as above.
56. Backpacks, Duffel Bags: these are useful in carrying items. Backpacks slip easily on your back, and duffel bags are light-weight.
57. Fishing supplies/tools: some good fishing poles, lines, hooks etc will be helpful in boosting your food supply.
58. MREs: Meals Ready to Eat, these are the current form of US military field food rations; what were once known as C-Rations. They come in a small box, and are a complete, well-balanced meal - albeit not entirely tasty. In time of desperation, a few cases of these meals can keep one person alive for weeks - perhaps months. If you're going to be in the wilderness for a long time - and working to re-build your own society - these are an excellent food source to use until you get your farm going and start to hunt and catch your own food.
59. Rice - Beans - Wheat: all of these are good foods that have a long shelf life.
60. Canned Fruits, Veggies, Soups, Stews, etc: as with the items listed above, these are healthy foods that will last a long time.
61. Milk - powdered & condensed: fresh milk will spoil quickly without refrigeration. Both of these will last a long time, and neither needs refrigeration.
62. Popcorn, Peanut Butter, Nuts: all of these contain vital nutrients and also last a long time.
63. Dehydrated fruits, vegetables: again, healthy foods that will last a long time.
64. Graham crackers, saltines, pretzels. Trail mix/Jerky: these are good foods, loaded with nutrients, and will last

a long time.

65. Chocolate/Cocoa/Punch (water enhancers): good for adding flavor to water, and also healthy and will last a long time.

66. Honey, Syrups, white/ brown sugar: refined white sugar attracts all manner of insects, and is not very healthy. These are excellent alternatives, and honey can last for literally centuries!

67. Garlic, spices & vinegar, baking supplies: these will aid in keeping you healthy, and baking supplies will allow you to cook your own food.

68. Soy sauce, bouillon, gravy, soup base: more good basic foods that will keep you healthy, they are small and easy to transport, and they have a long shelf life.

69. Canned Salmon Fish: fish is very healthy, and salmon is one of the best foods for you. By bringing canned fish with you, it'll last a long time.

70. Vegetable Oil (for cooking): full of good nutrients, and it will last a long time.

71. Flour, yeast, salt: all are needed for cooking, and will last - just be sure to keep them in tightly sealed containers to prevent insects and vermin from getting into them.

72. Teas & Coffee: both are good drinks to have.

73. Chewing gum/candies: gum is a good means of cleaning your teeth in lieu of toothpaste, and it also exercises your jaw. In addition, chewing gum is a good appetite suppressant. In times of short supplies, this can be critical. Candies are nice treats for the family, and provide a sugar boost.

74. Vitamins, Minerals: all are vital to good health. Have bottles of daily multi-vitamins on hand.

75. Medications: if any member of your family has any sort of health issues - whether it's asthma, hay fever, diabetes etc - make sure you have at least a three month supply of any drugs they need.

76. Sweatshirts/pants: in cold weather, these are important for being able to be active and stay warm.

77. Hats & cotton neckerchiefs: the sun can be very harsh. A hat is important for protection, and cotton neckerchiefs will keep the sun off your neck and cotton breathes.

78. Gloves (work/warming/gardening, etc.): you're going to be doing a great deal of physical activity, and you need

to protect your hands. Any injury can lead to you being unable to work, and this can endanger your family.

79. Socks, Underwear, T-shirts, etc. (extras): you have no idea when you'll be able to get new clothes. Keep plenty of everything on hand.

80. Woolen clothing, scarves/ear-muffs/mittens: if the weather is going to turn cold, make sure you have plenty of warm clothes.

81. Thermal underwear (Tops & Bottoms): same as above; you need to stay warm in bed when the weather turns cold.

82. Work boots, belts, denim & durable shirts: these go along with the work gloves; you need good, solid clothes that will last a long time and allow you to work hard.

83. Rain gear, rubberized boots, etc.: if you're going to be out in the wilderness for any length of time, you're going to need rain gear.

84. Mosquito coils/repellents, sprays/creams: in the warmer months, insects are going to be a pest, and you're going to want some relief. Bring as much of these items as you can. Not only are they small and inexpensive, but you can also trade them with others.

85. Mousetraps, Ant traps & cockroach magnets: you're also going to need to keep pests and vermin away from your living space and food - these items will do that.

86. Screen Patches and glue: without air conditioning, you're going to want to keep your windows open as much as possible - during the summer months. To keep insects and pests out, you'll need screens, and you'll also need to patch any holes that appear. So, get the materials and tools needed.

87. Nails, nuts, bolts, screws, etc.: good, all-purpose items needed for all manner of things.

88. Roll-on Window Insulation Kit: if you don't have storm windows for winter, this is a cheap and easy way to make your windows tighter for winter.

89. Cigarettes, cigars, pipes and tobacco: while not healthy (and you may not even smoke), these can be traded with others.

90. Wine/Liquors: even if you don't drink, these can be used to bribe people, trade with others, and even medical uses. If you need a quick antiseptic or anesthetic, these are ideal.

91. Lumber (all types): you'll need this for all manner of

activities.

92. Wagons & carts: helpful for transporting all kinds of items to and from your safe house.
93. Seasoned Firewood: new wood is green and does not burn well; you'll want wood that has been seasoned for at least a couple months to burn.
94. Bicycles and their supplies: these are excellent means of quick transportation. They don't need fuel or lubricants, and they're good exercise. Be sure to have tires, tubes, pumps, extra chains etc on hand.
95. Big dogs: not only are they good companions, but they can guard your home and warn you of the approach of danger. Be sure to have plenty of dog food. In a crisis situation, you can also (regrettably) eat them.
96. Cats: also good pets, and they can keep mice, rats, and other vermin away from your food supplies. Also, in times of shortage, you can eat them.
97. Goats & Chickens: these are ideal farm animals. Goats can live on just about anything, the females give a good amount of milk, they're small and easy to transport, and you can eat them. Chickens give eggs, can live on worms, insects and natural feed, and can also be eaten. Also, they can serve as a simple basic alarm system. At night, if someone tries to sneak into the area, they'll make noise if they're spooked.
98. Sundial, windup clocks, and perpetual calendar: keeping track of time is going to be critical to your survival, and you can't count on your watch, cellphone or any other electronic timepiece that needs batteries. So, a sundial set up outside and clocks that just need to be wound will help you keep track of time. Perpetual calendars are usually small metal devices that you turn a dial on to set the month and year, and then it shows you how the days of the month will line up. In the event of a long term disaster, you're going to need to keep track of the days of the years. There's a reason ancient civilizations like the Mayans and the Celts built Tulum and Stonehenge; these places marked the passage of dates important to their agrarian culture. If you're going to be reduced to living that way, dates like the summer and winter solstice, and spring and fall equinox are going to be equally important in planning your planting and harvest times.
99. Musical instruments: even if you and your family don't

know how to play any instruments, bring some simple ones along. A guitar, recorder etc can help (as with the games) to boost your morale and preserve your sanity. People need "nourishment" for their hearts and souls as much as (if not more than) their bodies.

100. Kindle, Nook and other types of ebook readers: a single ebook device can contain hundreds of books, and is thus an excellent way of keeping massive amounts of information and literature at your fingertips without all the weight of actual books. However, one word of warning here: *only* use such devices if you have a reliable power source to keep the batteries charged.

SURVIVAL TOOLS

While the Survival Kit refers to the portable basic items that are useful during short-term survival, the Survival Tools are the items that will increase your chances of surviving and comfort during long term survival.

-- LIST OF TOOLS & DESCRIPTION --

1. Canning supplies, (jars/lids/wax): once you start to harvest your own foods, you'll need a way to preserve them for later.
2. Knives & sharpening tools: files, stones, steel: even the best knives get dull with time. So, have the materials to sharpen them.
3. Bow saws, axes and hatchets, and wedges and honing oil: in order to hunt, cut down trees, make a shelter, split timbers for lumber and/or fire wood, all of these items are needed. Also, a small hatchet can be a weapon and used to slaughter livestock.
4. Garden tools & supplies: if you're going to be at the safe house for a long period, you're going to have to plant a garden to maintain your food supply.
5. Hand pumps & siphons: without power, getting water and fuels out of large containers (or abandoned vehicles and buildings) will be difficult. Also, you don't want to waste fuel on a power-driven pump.
6. Solar power systems: in the event of a long term survival situation you are not going to be able to count on a gas generator - you're going to run out of fuel. So, invest in some solar panels, batteries, and a complete installation system. A small unit can be bought for as little as a couple hundred dollars. While the system will not provide a large amount of power, it can be enough to keep a fridge going, a radio operating, and several other appliances working.
7. Windmills: these have multiple uses. They can pump water, provide electric power, and even turn fans to help you stay cool.

AIR, WATER, FOOD, FIRE, SHELTER

When preparing for a disaster, it's best to think first about the basics you're going to need to survive: clean air, fresh water, healthy food, fire, and shelter. As soon as all five elements are secured, you and your family will notice increased morale, and your stress will be relieved.

AIR
Of the five elements, this is the most critical. The average person can only hold their breath for about three minutes. Most of the speculated global disasters will cause polluted or poisonous atmosphere that, in some cases, could last for months or even years. It's crucial to prepare air filters and gas masks to be able to breathe during your search for a clean air supply. In the event of intense volcanic eruptions, or a nuclear, biological or chemical disaster, the ability to filter the air you breathe, and protect your body from contaminants will be vital.

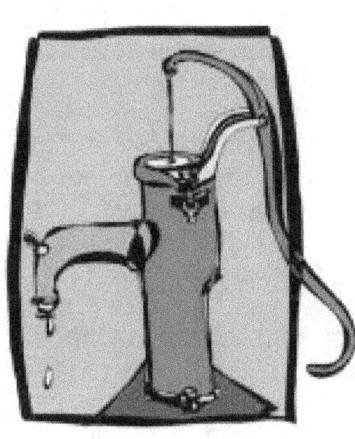

WATER
This is the second critical supply element. Remember that rule: three minutes without air, three *days* without water. So, maintaining a steady supply of water is vital. Under normal conditions the average person needs to drink at least two quarts of water every day. During a stressful, active fight for survival, especially in hot environments, your body can need *double* that amount. Water is also needed for food preparation and hygiene. At any time, during the initial evacuation and after you've reached your safe location, you should store at least one gallon per person, per day, and refill a full 7 day water supply at any possible occasion. If possible, have a two-week supply of water for everyone in your family.

In an extreme situation, when you're running low on your supply of water, try to *not* limit your drinking water rations. It is vitally important that you drink enough water each day; otherwise heat stroke, dehydration, kidney troubles, and a host of medical problems can occur. Instead, focus on finding a new water supply. You can minimize the amount of water your body needs by reducing your physical activities and staying cool.

Also, forgo bathing and even brushing your teeth. In a critical situation, even one ounce of water can be the difference between life and death. Keep in mind that bathing does *not* require water that is completely clean - what is known as *potable* water - meaning water fit to drink. You can make use of a simple mountain pond to bathe - if necessary.

FOOD
Preparing for a scheduled evacuation will give you enough time to organize a long term "shelf-stable" food supply that will be sufficient to survive for months. However, the MRE meals and canned foods currently available on the market will disappear from the shelves in just a few days after a global disaster is announced. It's important to prepare at least a 1 month supply of high calorie shelf-stable food per person that you can take in the event of either an immediate or scheduled evacuation.

During your long term survival after the disaster, you'll have enough time to teach yourself the most efficient ways to gather or hunt for food. Remember - do not take any frozen food, as this will only waste space during transportation, and will be useless when it spoils. As for fresh food, some fruit during your initial trip can be good, but don't bring too much. It will soon spoil.

FIRE

Staying warm and dry during long storms and extremely cold winters is also vital, as a human can suffer from hypothermia in just a few hours being exposed to extreme cold, especially when exhausted and starving. A permanent source of fire, such as a magnet-steel fire starter is a must-have item in every Survival Kit. It's important to find a shelter near a source of firewood or any other heat producing materials. If you're lucky, you might have a supply of heating oil, coal, gasoline, or even peat. The latter is a thick organic material (essentially the precursor to coal), and burns quite well - albeit with a strong odor.

Also, fire is useful in preparing warm meals and drinks. In an extreme situation, you can forgo using fire for any sort of food preparation. Many MREs can be eaten cold, and there are plenty of drinks that do not require heating. While doing without your morning cup of hot coffee might be inconvenient, in a life or death situation it is a reasonable sacrifice.

SHELTER

This is the final aspect of staying safe and staying alive following a major disaster. In the case of a truly global cataclysm, adequate shelter is one of the most important lines of defense to ensure the safety of you and your family. Having a proper shelter will protect your family during the initial aspects of the disaster (fire, flood, polluted air etc), and then give you a decent starting point upon which to re-build your lives after a natural or manmade disaster.

No matter what sort of disaster befalls your area (or the entire world), one thing is certain: there will be widespread civil unrest, riots, looting, and possibly Martial Law. It will be important that you have a safe and secure shelter in which to store your food, water and other supplies, and to keep your loved ones safe.

PERSONAL PROTECTION
This is a very sensitive subject, as it deals with the possibility of causing harm to other people, and (by and large) most people don't want to do that. Sure, there are all kinds of action movies and TV shows out there that portray exciting action, but - the reality of it is - violence, knives, guns, and so on are not fun, not pretty, and not at all pleasant. However, in the event of a major disaster, you are going to have to protect yourself, your family and your essential supplies.

If you're reading this book, then that means you're committed to surviving, and that means preparing for the worst. On the other hand, there will be thousands (if not millions) of people who did *not* have the foresight to prepare for the flood, fire etc. You've seen these people, on the news, when word of a hurricane or some other disaster is announced. What do they do? Some bolt to the stores and buy up anything and everything they can get their hands on - even things as stupid as TVs, stereos, DVDs, and so on. Others engage in general lawlessness - trashing cars, stores etc, stealing clothes, setting fires, raping, and killing others.

Not the sorts of things you want your children and loved ones exposed to. Once the general chaos has subsided (and the looters have dropped their big screen TVs by the wayside), these same people are going to want water and food, and they are going to use *any* means necessary to ensure their own well being. Some of these people will be strangers to you, and they will commit any crime they want in order to get what they think they need. In other cases, they could be your next door neighbors, co-workers, and even trusted friends. Unless they're part of your evacuation group - as sad as it may be to think of - you'll have to leave them behind, when you leave for your bunker/safe house. Remember, you have only food, water and supplies for your group.

Although not a pleasant thought, you must keep your priorities in order and do *whatever* is necessary to protect yourself, your family, and your ability to survive. That means securing weapons - whatever you're comfortable using - guns, knives, rifles etc and learning how to use them effectively. If necessary, you may have to kill to survive - remember that! If you feel the least bit hesitant, just think about your spouse and loved ones being raped, tortured and killed before your eyes.

A key element to remember is this: a simple leg wound will effectively stop an attacker and allow you to flee the scene. If you can't bring yourself to kill, use a crippling wound to at least protect yourself and your family.

TEMORARY SHELTERS

Depending on the type of disaster that befalls us, you may only have time to make use of a temporary shelter. Also, in the event you're far from home (away on a business trip, on vacation etc) when a disaster occurs, you may not be able to get to your home or Safe House. In that case, some type of temporary shelter will have to do. The type you get will be a function of several factors:

- Location: where you are at the time of the disaster - the ocean, a city/urban setting, a desert, jungle etc.
- Climate/Time of Year: are you in the north or is it winter?
- Event: what sort of disaster has happened? A flood, meteor impact etc.
- How long will it take for you to build a shelter, and how much work will be required?
- What sort of weather conditions (wind, rain, heat, cold) will you need protection from?
- Do you have the equipment/tools needed to build the shelter?
- If you don't have them, can you fabricate the tools?
- Do you have the proper materials to make the shelter, and do you have enough of the items?

In order to answer the above questions, you have to know the steps in making the different forms of shelters, and you'll also need to know what items you have to have to build them. So, let's look at each shelter.

Figure 2. Poncho Lean-To.

Poncho Lean-To:

This is a quick and easy shelter to make. All you need are the following:

- A poncho
- Some rope, about two to three meters in length
- Three stakes, each about 30 centimeters in length
- Two poles or trees that are two to three 3 meters apart

Now, this is important, prior to selecting the location for this shelter, figure out which way the wind is blowing. You want to make sure the back of the lean-to is facing the wind.

Construction Steps:

1. Tie the poncho's hood off, and then pull on the drawstring until it's tight. Roll up the poncho lengthwise, and then fold it into thirds. Then use the drawstring to tie it off.

2. Take the rope and cut it in half. Take each piece and tie them to a corner grommet of the poncho.
3. Tie a short stick, approximately ten centimeters in length, to each of the ropes, and position each stick a couple centimeters from the poncho. These are drip sticks, and they'll prevent rainwater from trickling into your lean-to from the ropes.
4. Cut some lengths of string, about ten centimeters long, and tie them to the grommets along the top edge of the poncho. These will keep water from dripping into the shelter from the top edge.
5. Run the ropes to the trees, and tie them at about waist high.
6. Spread out the poncho and then secure it down to the ground using some sharpened sticks. Run the sticks through the poncho's grommets, and down into the ground.

If you need to sleep in the lean-to for several nights, or the sky gives indications of rain, you'll need to add a center support to the lean-to for added strength. Just take a length of rope or line, tie it to the hood of the poncho, and then run it up to any branch hanging over the lean-to. Pull the line taut and tug the branch down a bit, and then tie it off. This will ensure that the line has no slack in it.

You can also use a short branch or stick to support the lean-to's center point. However, if you use this method, the stick will partially block the opening. So you'll need to be careful as you climb in and out of the shelter.

If the area you've set up the lean-to in is prone to rain and wind, you can add protection by putting some brush, your backpack, or anything else large and heavy along the sides of the shelter.

You also want to keep in mind that, at night, the environment is going to cool off, and while you're asleep you can lose up to 80% of your body's heat into the ground. So, some sort of insulation is important. Look around the area; are there leaves, pine needles, or do you have some clothing items you can spread out on the ground to sleep on?

Finally, just how dangerous is your environment? Specifically, are you trying to hide from people? Depending on the disaster that's occurred in your community, you may need to maintain a low profile to stay safe from possible discovery. To help you achieve this, you can give the lean-to a low profile - literally. Instead of tying the support ropes waist high to the trees, put them down at your knee height, and then angle the poncho so that it is down toward the ground. You can then use sharp sticks to secure it. The lean-to will be harder to get in and out of, but this will give the lean-to a lower silhouette, and thus make it harder for people to spot.

Figure 3. Poncho Tent.

Poncho Tent:

This is yet another means of creating a tent from a poncho. It has the virtue of having a low silhouette, which can be important if you're in a dangerous area. While it gives you protection from the weather on two sides, the tent has less space than a lean-to.

To build it, you'll need the following:

- A poncho
- Two ropes, each about 1.5 to 2.5-meters long
- Six sticks, each about 30 centimeters long, and sharpened
- Two trees, and they need to be 2 to 3 meters apart

Construction Steps:

1. Tie the poncho hood off the same way as you would to make the poncho lean-to
2. Tie one of the ropes to the center poncho grommet on either side
3. Run the ropes to the two trees, and tie them at close to knee height
4. Stretch out the poncho until it's tight
5. Pull one of the poncho's sides down to the ground, keeping it tight, and push some of the sharpened sticks into the ground through the grommets
6. Use the identical procedure to secure the other side of the poncho

If you want a center support for the tent, you can use one of the methods outlined above for the poncho lean-to.

As an alternative center support, you can make an A-frame, which is set outside the tent, over its center. Take two sticks, each about 90 to 120-centimeters long, and make sure that one stick has a forked end; this is what will make the A-frame. Tie the ends of the sticks together, bend the forked end over (be careful not to break it), and secure it at about the center point of the other stick. After that, pull the drawstring of the hood up and tie it to the A-frame. This will give good support to the tent's center.

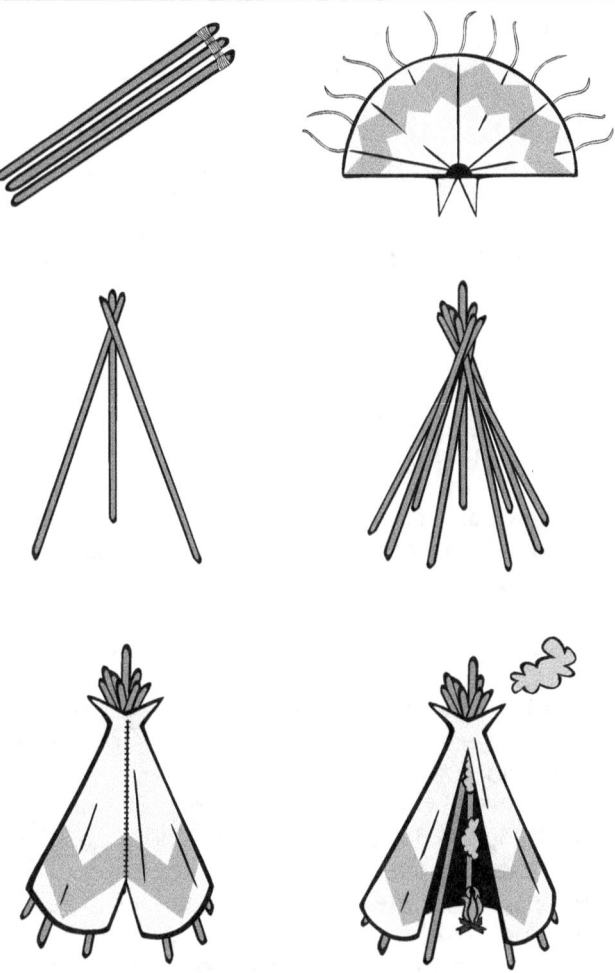

Figure 4. Three-Pole Parachute Tepee.

Three-Pole Parachute Tepee:

As its name implies, you need a parachute to make this kind of tent, along with three poles. If you have these items, a parachute tent/tepee takes little time to construct and is easy to make. Tents made from a parachute have several pluses:

- Providing protection from harsh weather
- They can be used for a signaling device, especially if you use a flashlight, candle or fire to enhance its brightness
- The tent will be big enough to accommodate a couple of

people and their gear, and still have plenty of room to
sleep, cook, and store supplies

This tepee/tent can be built using an entire parachute or
just parts of one, and it works with either a main 'chute or a
reserve. If you have a standard parachute as used by individual
military personnel or skydivers, you'll need three poles; each
will need to be between 3.5 to 4.5 meters long, and around 5
cm in diameter.

Construction Steps:

1. Put the poles on the ground so that their ends are close
   together, and then tie those ends together with a stout
   rope.
2. Stand the poles up, being careful to not let them fall,
   and then spread them out in the form of a tripod.
3. For better support, put more poles against the three
   making up the tripod. Any number will help, but five
   work best. Also, don't tie the additional poles to the
   tripod.
4. Figure out which way the wind is blowing, and then
   position the entrance for the tent 90 degrees (or more)
   from that direction.
5. Put the parachute on the "back" of the tripod, and get
   the nylon web loop (known as the bridle loop) positioned
   at the (apex) (top) of the canopy.
6. Put the web loop over one of the free-standing poles.
   Withdraw the pole and then put it back against the
   tripod to hold the chute in place. You want the apex
   of the canopy up at the same position as the rope tied
   around the tops of the three poles.
7. Wrap the chute around one section of the tripod;
   make sure you have the chute doubled up (for added
   thickness). Once the chute is in position over half of
   the tripod, wrap the remainder around the tripod in the
   other direction.
8. Create an opening by putting the folded edges of the
   chute around two of the free-standing poles. Then, any
   time you want to close the entrance, simply move those
   poles so that they're side by side.
9. Any extra bits of the chute can be tucked under the

poles and then spread out inside the teepee to make a floor for your comfort.

10. Be sure to leave an opening at the top of the shelter, about 30 to 50 cm in size for ventilation, and to vent any smoke if you build a fire inside.

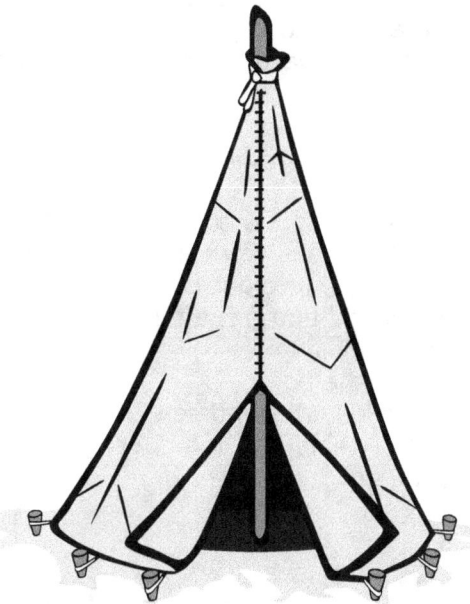

Figure 5. One-Pole Parachute Tepee.

One-Pole Parachute (or similar tarp) Tepee:

In the event you're rather short on supplies, you can still make a shelter from a parachute. To construct this tent, you'll need the following:

- One 14-gore section of a parachute
- Some short stakes
- A stout center pole
- An inner core of a chute and needle

Cut the chute's suspension lines off, except for those on the chute's lower lateral band. Leave those at 40 to 45 cm lengths.

Construction Steps:

1. Select a site for the shelter and draw a circle having a diameter of about 4 meters on the ground.
2. Use the stakes to tie down the parachute using the lines on the lower lateral band.
3. Select a spot for the door, place a stake, and then tie the first of the lower lateral lines to it.
4. Stretch out the parachute toward the next line until it is taut, then place another stake on the circle, and then tie the line to the stake.
5. Continue this process until all of the lines have been tied down.
6. Attach the top of the chute loosely to the center pole. Use one of the suspension lines you cut beforehand, and then, using trial and error, figure out the point where the parachute will be pulled up tight after you've put the center pole in its upright position.
7. Secure the parachute to the pole.
8. Take one of the suspension lines and sew the end gores together; just leave an opening for the door (generally a one meter opening will do).

Figure 6. No-Pole Parachute Tepee.

No-Pole Parachute (or similar tarp / large sheet material) Tepee:

In the event you don't even have a pole to act as a center support, you can still use the same material to create a shelter.

Construction Steps:

1. Take a rope (use one of the suspension lines you cut from the chute) and tie it to the top of the parachute.
2. Toss the rope over a stout tree branch, pull it down, and tie it off at the tree's trunk, but do not pull the chute up too high.
3. Working from the side opposite the door, place a stake into the circle you outlined on the ground.
4. Tie one of the lines from the lower lateral band to the stake.
5. Continue this process - pushing stakes into the ground and tying lines to each of them.

6. Once the parachute is completely staked out, loosen the rope from the tree trunk, and then pull on the rope to bring the chute material up. Once it is tightened, re-tie the rope to the trunk.

Figure 7. One-Man Shelter.

One-Man Shelter:

In a pinch, if you just want to build a quick shelter for just yourself, you can build a one-man shelter with a parachute, three poles, and a tree. Make sure that one of the poles is about 4.5 meters in length, and the others are about 3 meters.

Construction Steps:

1. Take the longer pole and tie it to a strong tall tree at about your waist height.
2. Put the other two poles on the ground so they're pointing in the same direction as the main pole, and put

one pole on either side of the main pole.
3. Take the parachute and lay its folded canopy over the main pole. Make sure that close to the same amount of the chute is hanging over each side.
4. If there's any excess material, tuck it under the shorter poles, and then spread it out on the ground on the inside of the shelter to create the floor.
5. Use stakes or some sort of spreader between the shorter poles right at the entrance; this will keep the poles from sliding inward.
6. If there's any excess chute material, use it to cover up the shelter's entrance.

Parachute material is very strong and durable; so the shelter will be wind resistant, however any kind of large durable sheet material can be used for shelters construction. In addition, the shelter is so small that it can easily be warmed. This may sound amazing, but you can use a *candle*, if positioned properly, to keep the inside of the shelter a comfortable temperature! However, extreme weather - like a snow storm - will damage the shelter, causing it to cave in.

Figure 8. Parachutee Hammock.

Parachute Hammock:

If you have 6 to 8 gores (sections) of a parachute, you can easily string it up as a hammock. All you need are two sturdy trees placed about 4.5 meters away from each other.

Figure 9. Fire Reflecting Field-Expedient Lean-To.

Field-Expedient Lean-To:

When you're in a forest/wooded area, and there are plenty of natural materials available, you can build one of these lean-tos without using any tools at all, or by just using a knife. This kind of shelter takes a while to create (or at least more time than many other kinds of shelters), but it is very good for protecting you from the weather. Here's what you'll need to build it:

- Two trees that are separated by about 2 meters of open space. If you can't find two decent trees, you can use upright poles.
- One pole that's about 2 meters in length and has a diameter of 2.5.
- 5 to 8 poles with the same diameter as listed above, and that are about 3 meters in length.
- Some rope, vines or cords to secure the horizontal support in place.
- Some saplings, rope or vines to crisscross the beams.

Construction Steps:

1. Take the 2-meter long pole and tie it to the trees, making sure that it is between your waist to chest height above the ground. This will serve as the shelter's horizontal support. If you have only one standing tree (or none), you can make a biped support for the support with Y-shaped stick(s) or tripods.
2. Take the 3-meter poles and stand them on the ground so that the other end of them leans on the horizontal support. Make sure the backside of the lean-to is toward the wind.
3. Take some of the rope, vines or saplings and crisscross them over the beams.
4. Get some pine needles, grass, leaves, brush etc and cover the framework. The best way to do this is like adding shingles to a home: start at the bottom and work your way up.
5. For bedding, get some pine needles, straw, grass or leaves and spread them around inside the lean-to.

If you're in particularly cold weather, you can improve the comfort of your lean-to by making a fire reflector wall. To build it, use the following steps as a guide:

- Get four stakes, all about 1.5-meter-long and drive them into the ground.
- Stack green logs between the stakes to create two rows with an inner open space.
- Fill the space with dirt. This will strengthen the wall and make it more reflective to heat.
- Use vines, rope or other material to bind the support stakes together at the top; this will keep the logs and dirt in place.
- You then build your fire in front of this wall, and it will reflect heat back toward you and the interior of the shelter.

Figure 10. Swamp Bed.

Swamp Bed:

If you find yourself in a swamp or marsh situation, or an area that has wet soggy ground or standing water, a swamp bed is ideal for keeping you high and dry. When you're looking for the best site to build the bed, here are the things to consider:

- The weather
- What sort of wind you can expect
- Is the water tidal, and if so, what sort of tide can you expect
- What materials are available to make the bed
- Here's what you'll need to build the bed:
- Four trees in a roughly rectangular arrangement, or four poles you cut to use
- Some rope or stout vines
- Two poles long enough to reach across the full width of the rectangle
- A number of shorter poles to span the length of the rectangle
- Bedding material: grass, leaves etc
- Clay, mud or silt to make a fire pad

Construction Steps:

1. Find four trees that form a roughly rectangular layout. If you can't find such a layout, cut four stout poles (bamboo is a perfect material) and pound them into the ground deep enough to give good support, and lay them out in a rectangle format. They need to be far enough apart to make a bed big enough for you to lie on, and have room for a fire and your gear. They also need to be strong enough to support your weight, and the weight of your equipment.
2. Take the two long poles (the ones long enough to span the rectangle's width), and tie them in position. Be sure they're strong enough to also support the combined weight of you and your gear. Make sure the poles are positioned high enough to keep you out of the water. If the water is tidal, make sure the bed is above the high water mark. You can determine this by looking for water stains on trees, rocks and other formations. If the water is from an ocean, there may be seaweed and/or barnacles at the high tide mark.
3. Take the other poles and use them to span the length of the rectangle. Place them on the two side poles, and then secure them in place.
4. Use grass, leaves, moss or any other comfortable material to cover the bed and create a soft surface to sleep on.
5. To create a fire pad, place a layer of silt, clay, or mud at one end of the bed. Allow to dry, and then build a fire on it to help you stay warm.

A variation of the swamp bed is useful when you don't have any rope or other material to tie down the poles. You use the same rectangle-like layout, but then you just place sticks, branches and other materials within the trees, or poles, until the sleeping deck is up high enough to be above the water.

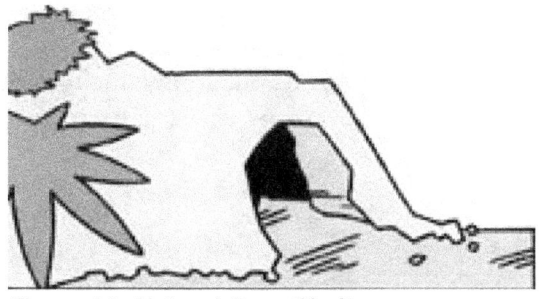

Figure 11. Natural Cave Shelter.

Natural Shelters:

Now, in the event you have no building supplies, you can seek shelter in any sort of natural formations. Here are just a few places that provide natural shelters:

- Caves
- Rocky crevices
- Clumps of bushes
- Small depressions in the ground
- Large rocks located on the leeward (protected) side(s) of a hill
- Large trees that have low-hanging limbs
- Fallen trees with thick branches

While all of these can offer you shelter, there are some things to consider. When you're looking at a natural shelter, do not stay in any low areas like ravines, creek beds, or narrow valleys. At night, low regions are where the cold night air, which is heavier than warm air, collects. As a result you can freeze there. Also, if it rains, the runoff will likewise collect in the low areas. This can lead to your campsite being washed out, and possibly you drowning.

Another factor to consider is avoiding ground that has thick brush; this type of environment is where insects love to collect. Before selecting a natural location as a shelter, check it over for poisonous snakes, mites, ticks, scorpions, and ants. While most ants are harmless, there are several varieties (like fire ants) that can sting.

Finally, check the area around where you want to camp for dead tree limbs, loose rocks, and any sort of vegetation (coconuts, fruit etc) that might be above you. If any such items fell on you, you could be injured.

Figure 12. Debris Hut.

Debris Hut:

If you're in a situation where staying warm is critical, and you need a shelter that's easy to construct, a debris hut is ideal. When you need a shelter because your survival is in question, building this type of shelter is your best bet to staying alive. Here are the materials you'll need:

- A long stout pole to serve as a ridgepole
- Two to four short stakes (if one or two trees are not available)

- Rope, vines or some other type of line
- A number of large sticks
- A number of small sticks and/or brush
- Insulating material such as pine needles, grass, leaves

Construction Steps:

1. Place the ridgepole in the branches of two adjoining trees at about waist height, and tie in place. If trees are not available, use short stakes to make tripods to hold the ridgepole in place.
2. Take the large sticks and place them along the ridgepole on both sides. Your goal is to make a triangle-shaped shelter under the ridgepole. You have two key aspects to be aware of. One, the area under the pole needs to be wide enough so you can comfortably lie down there, and two, the sticks need to be placed at a steep enough angle that rainwater and dew will run off.
3. Use the smaller sticks and/or brush to cover the larger sticks, placing them crosswise so as to create a lattice-like framework. This will prevent the insulation material you spread across the shelter from falling in on you while you're sleeping.
4. Spread any dry, light soft debris (grass, pine needles etc) across the ribbing. If possible, pile the material up until it's about 1 meter (3 feet) thick; the thicker you can get it, the better.
5. Spread out some of the insulating material inside the shelter for you to lie on. Again, the thicker you can put it, the better your level of comfort will be.
6. If you don't have any material to build a door, just put some of the insulating material in a pile by the entrance. Then, once you get inside the shelter, you can close up the opening by just pulling the pile toward you.
7. If the area you're in is particularly windy, you may want to add some branches or some kind of shingling material to the top of the shelter to keep the insulation from blowing off.

Figure 13. Tree-Pit Snow Shelter.

Tree-Pit Snow Shelter:

This is the sort of shelter to build if you're in a cold, region with lots of snow. To build it, you just need a shovel (or some other tool to dig with), a knife of some sort, and some evergreen trees.

Construction Steps:

1. Located a tree that has big bushy branches, these will provide cover over your head.
2. Using a shovel or other equipment, excavate some snow from around the tree trunk right below the low-hanging branches. You want to either get down to the ground level or a depth suitable to your needs. As for the diameter of the hole, it needs to be big enough to accommodate your body with ease.
3. Gather snow from around the area, and pack it tight inside the hole and around the top; this will give your body a good amount of support.
4. Using a knife, cut some bushy branches from some

other evergreen. Put some of them across the top of the pit (they'll provide more cover), and put some others in the pit (they'll provide insulation).

Figure 14. Beach Shade Shelter.

Beach Shade Shelter:

On the other hand, if you find yourself on a beach (maybe you're shipwrecked or stranded on an island), this type of shelter can protect you from wind, sunlight, rain, and heat. You can make it very easily, and all you need are the following natural materials:

- Some long flat pieces of driftwood
- A short piece of driftwood that can act as a shovel
- Grass, leaves or other soft material, but not seaweed! This is your bedding material, and seaweed will have a bad aroma

Construction Steps:

1. Gather together as much driftwood and/or other kinds of natural materials. These will be your shovel and roof supports.
2. Find a site for the shelter, and make sure that it is above the level of high tide. You can determine the high water mark by looking at where seaweed comes up on the beach.
3. Use your shovel tool to dig a trench in the sand. Make the trench in a straight line going north to south; this will ensure you get the smallest amount of direct sunlight. You want the trench wide enough for you to lie down in and roll around comfortably, and long enough to accommodate your height. To determine where north and south are, see where the sun rises in the morning -

that's east. If you stand facing east, north is to your left and south is to your right.

4. Do not dig down too deep; groundwater will seep into the trench. Instead, scrape up sand around the trench on the two long sides and one of the short sides. The more you pile the sand up, the more headroom you'll have in the shelter.
5. Put the driftwood (or other materials) across the trench to create a roof. If you have access to any rocks or heavy objects, use them to give the roof weight to prevent it being blown away.
6. Use the shovel to dig out the entrance to the shelter; make is large enough to allow you easy access to the interior.
7. Place materials like leaves, grass etc inside the shelter to create a bed.

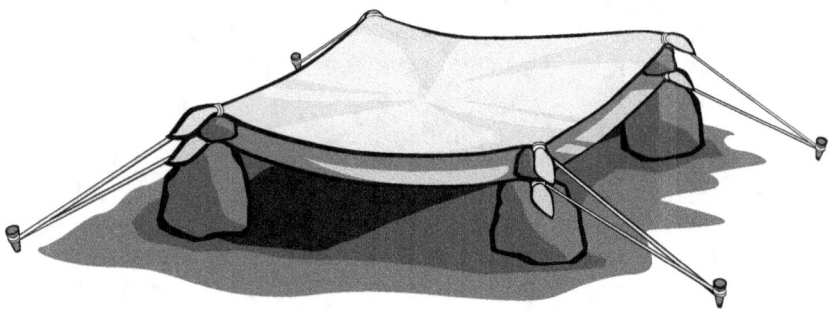

Figure 15. Open Desert Shelter.

Desert Shelters:

If you find yourself stuck in a desert situation, you need to consider the work, time, and materials you'll need to make a shelter. Remember, any degree of physical labor will increase your dehydration and need for water. So, try to find something as simple as possible to build.

To that end, try making use of something like a piece of canvas, a poncho, or a parachute, and spread it over a natural terrain feature. Look for an outcropping of rocks, some sand mound, or a depression located between rocks or dunes. This is a simple, basic shelter that you can build in very little time and with little physical exertion.

If you have a rock outcropping to use, here are the construction steps:

1. Use some weighted objects (rocks or other items) to anchor an end of the canvas, poncho, or whatever material you have available to an edge of the outcropping.
2. Stretch out the canvas as far as you can to another portion of the outcropping, and similarly secure it in place. Your goal here is to provide the most shade possible.
3. If you're in an area that's sandy, use this method:
4. If a sand dune is available, use it. If not, build up a mound of sand with a shovel, tool, or even just your hands.
5. Place an end of your canvas (or other item) on top of the mound, and then anchor it. If you have a rock, use that. If nothing is available, merely stuffing the corner of the material into the sand can work.
6. Stretch out the canvas and secure it in a similar manner. As with the above process, your aim here is to provide shade.
7. Note: If the material you have is big enough, fold it in half. An airspace can further reduce the temperature in your shelter.
8. Another means of reducing the level of heat in your shelter is to make one that goes belowground. However, to build such a shelter will require more time and physical exertion, which means more sweating and dehydration. So, if you're short on water, consider forgoing this structure.

Construction Steps:

1. Locate a low area or depression between rocks or dunes. If you need to, dig down about 45 to 60 cm and create a trench. Make it wide enough and long enough that you can stretch out in it comfortably.
2. Take the sand from the trench and use it to build a mound on three sides of the trench.
3. At the other end of the trench, you'll need to dig down a bit more to create an entrance.
4. Take the canvas, sheet or whatever material you have available, and use it to cover the trench.

5. Use rocks, sand or some other weighted materials to secure the cover in place.
6. As with the shelter listed about, any extra shelter material can be folded over to help reduce the midday temperature inside.

Figure 16. Snow/Ice Shelter (Igloo).

Snow/Ice Shelter (Igloo):

On the flip side, if you're facing winter conditions and need to build a shelter out of essentially nothing, an igloo is the ideal solution. All you need to build one is some type of knife or cutting device. There are seven criteria to follow to insure you build a good igloo:

1. <u>Shape</u>: make the igloo round; this is because the arch is one of the best geometric shapes in terms of strength. So, you want the igloo round, and as smooth as possible - to reduce wear due to the wind.
2. <u>Size</u>: you want the igloo as small as possible. This is for two reasons. First, a small shelter is easier to build, and second, it's easier to heat.
3. <u>Venting</u>: this is something most people don't think of, because they don't realize that snow does not breathe. You need to cut a small hole (about the size of a softball) in the side of the igloo. Put the vent about halfway up the wall, and at a 45 degree angle. Putting it over the door is best; that way you won't get snow falling on you while you sleep. A good way of keeping the hole clear is to put a branch, stick, ski pole etc in it,

and then shake it every once in a while during a snow storm.

4. Air Sump: cold air always drops to the lowest point in a room. As a result, you need to create a sump (low spot) in the igloo below the level of your bed where the cold air can collect.

5. Door Size and Shape: you want it as small as possible to minimize heat loss from the igloo and cold air intrusion, so you don't need a lot of material to block it while you're inside. Ideally, make it just big enough for you to crawl through, and small enough so a backpack can cover it. As for size, make it with an arch - it is the ideal shape for maximizing strength to support it. When you cover the door, it is okay if the seal is not perfect - a little air getting in will help the igloo to vent.

6. Door Location and Position: check the wind direction while you're building the igloo and point the door away from the wind. You also want the door positioned so that it is lower than the sleeping area, which should be a high flat area. Inside the igloo, the cold air will settle to the sump and the warm air created by you cooking, running a heater, or just from your body temperature will rise. If the door is placed too high in the igloo, all that warm air could simply slip out.

7. Location: build the igloo out in an open area and away from any trees. When snow falls from the branches of trees onto the snow on the ground, the crystals within the snow collapse. As a result, the snow becomes hard and difficult to use as a building material. By using snow that has not had other snow fall on it; you get good snow that is ideal for building a shelter.

If you have the time and money to buy a shelter kit, there are many available to you. Even a simple internet search for "Temporary Shelter" will yield a plethora of companies that provide a host of such structures. First, let's consider some universal features to any shelter:

1. Make sure that it has the UL (Underwriters Laboratory) seal of approval.

2. Review the specifications on the unit to see what sorts of loads (wind loads, snow loads etc) it is rated to withstand.

3. Make sure the shelter is easy to construct. Ideally, you want a unit that requires few (if any) tools. Pieces should snap and lock together with ease.
4. Now, let's look at some features you'll want to consider for different conditions.

Desert

In this area, your key concerns will be heat and wind. You need a structure that will stay cool, and will be able to withstand potentially high winds (winds that will whip up sand that can rip through delicate cloth and flesh!). So, stay away from metal structures - they'll become hot - and stick with thick reinforced polyester that is treated to resist ultraviolet radiation. Many temporary shelters come with small air conditioning units, and this will be important to survive in a desert climate.

Jungle

While similar to the desert situation, the heat of a jungle is mild in comparison. As a result, a metal structure is preferred and will stand up better in protecting you from wild animals. A key element to consider is getting a unit that has a base you can easily secure to the ground or a foundation. Also, if you are unable to pour a concrete foundation, select a site that has a good solid base - ideally stone - and bolt the shelter down. Be sure to select a spot that is above any nearby river or body of water - in the event of flooding.

Winter

This is the flipside of the desert situation; you want a unit that is well insulated and rated to withstand the high winds associated with winter storms. A shelter can be made of PVC tubing and tough, tear-resistant, non-puncture tarp material. Most shelters are white - to allow natural light to filter inside, yet not be glary. The panels need to be UV-stabilized so they can withstand years of exposure. Granted, you are not expecting to use the shelter that long, but most are built that way. Also, get a shelter that comes with ventilation tubes. These can be installed between the panels, and will insure that the shelter does not overheat. That might sound silly in winter conditions, but on a clear day, and cooking inside, the shelter could grow hot, and thus venting to the outside will be important.

Ocean

There are two situations where you need an ocean-worthy shelter. The first is if you're on land and a tsunami is heading your way, and the second is a raft to carry you across a large body of water. For the former, you need a small, compact unit that can withstand the initial tidal impact and not be damaged as it is buffeted about during the aftermath. Also, it has to be padded inside to protect you from physical injury.

If you're going to be adrift for some time, a sturdy raft with a cover is critical. The cover is important to protect you from bad weather and sun exposure. Also, be sure the raft has a rainwater collector. Depending on how long you have to be in the raft, drinking water may become critical, and every drop of water you can collect from the rain could save your life.

UNDERGROUND LONG TERM SHELTER

In the event you're looking at a truly massive global disaster - Nuclear Winter, plague, asteroid impact, radiation fallout etc - a simple Safe House may not be enough. No, you may need an underground shelter that will be able to sustain you and your family for a long time. You're going to need it to contain all of the components vital to survival:

- Food
- Water
- Air
- Waste disposal

In essence, the ideal shelter would be a submarine! This might seem odd, but - if you think about it - that is what a long term shelter needs to be. There are numerous companies that sell just such units. Depending on the size of your family, you can get underground shelters that have a central living room with rooms attached in a sort of circle about the hub. There will be sleeping areas, storage rooms, air processors, waste disposal, a kitchen and eating area, and even a recreational room. How big and how fancy you get depends on your budget, time available to construct the unit, and the size of the land for your shelter. These can run upwards of $80,000. Some companies will even deliver the shelter fully stocked with supplies to enable your family to survive for up to two years! These shelters are water proof and air tight, and can be delivered and installed at night (to avoid being seen by anyone). Some companies even construct a sort of "bird blind" cover structure - a shed or other innocuous building to hide the entrance to the shelter. Thus, when disaster strikes, you and your family can enter the shelter in safety.

If your budget doesn't allow for the purchase of a major underground shelter, you can construct (either yourself or hire someone) one. These shelters can be made of cinder block, concrete, wood, stone, brick, or metal (steel is best). A key element to remember is that the deeper you place the shelter the stronger its walls and ceiling have to be to hold back the dirt. Also, the wider each room is the more roof supports (posts or poles running down the center of the room) you'll need. On the other hand, if you make the room too narrow, you could get feelings of claustrophobia over time.

While a stone, brick or cinder block shelter will be strongest, unless you know how to work with cement and masonry, you're going to have to hire someone to build your shelter. This will entail greater cost, and you'll have the added problem of someone who is *not* part of your family knowing where your shelter is located! When disaster strikes, that worker (or workers) may show up wanting to take over your shelter. Ask yourself: are you prepared to keep them out?

The cheapest and easiest type of shelter is a wood-framed one, and can be built for as little as a couple thousand dollars. Just about anybody can construct one, and plans are easily available from numerous sources. A quick online search will yield you a plethora of companies offering such plans. For that matter, there are even free schematics you can download.

One way of minimizing the size and strength of your shelter is to place it as shallow under the ground as possible. Typically, an underground shelter needs a minimum of three feet of cover (that is, three feet of dirt over its ceiling) to be completely safe. However, that is not vital - you can make do with less in order to save time and money in construction. Another reason to not place your shelter deep underground has to do with the location of groundwater. Water underground rises and drops over time, like an ocean tide going in and out: dropping during dry spells and rising during the rainy season. How high it gets depends on the environment you're in. If you're in a very wet region (say, Florida, for example) the groundwater might get all the way to the surface during the wet season, and swamps and marshes may form. On the other hand, if you're in a desert region, the groundwater won't be an issue during any period of the year.

There are reasons to be concerned about where the groundwater is located in the area you plan to build your shelter. First, there's weight - water makes the dirt heavier.

If you design your shelter to withstand the weight of *dry* soil, it could collapse once the groundwater rises above the height of the shelter's ceiling! Second, there's the issue of waste disposal. Wherever you position your toilet, the water (when you flush) will need to run "downhill" (so to speak). If the groundwater level is above the level of your toilet, it won't be able to flush - it'll back up into your shelter! This is yet another reason to locate your shelter not very deep underground (unless the groundwater is quite deep in your area).

A good way of figuring out how to position your shelter is this: first, find out how high the groundwater gets in your area during the wettest time of year. A geologist can help you determine this. Using that elevation, find out from a plumber at what elevation you need to set your toilet (and the floor elevation of the bathroom). From that, you can determine how high your ceiling has to be, and that will tell you how much cover the shelter will be able to have. In some cases, that could be several feet (remember, it only has to be three feet; more than that is "gravy" - as the saying goes). In other places, it could be virtually nothing.

Third, there's the matter of drinking water. Each person in your shelter is going to need at least a gallon of water each day. Depending on the size of your family, this could mean hundreds (maybe over a thousand) of gallons, if you have to stay underground for a year! If you're unable to store that much water, you're going to have to draw from the surrounding groundwater, and then purify it prior to use. Ideally, you'll want to position your sewer line dumping out into the ground and your water intake line as far away from each other as possible. Groundwater has a flow to it - much like a river or stream. So, place your drinking water line "upstream" of your sewer line, and that way you'll avoid taking in any of your sewage.

Food is the next critical element. You'll need to stock up on non-perishable items; MREs are the best things to get. They last a long time, have plenty of nutrients, are a well-balanced meal, and take up minimal room. Their flavor is said to be lacking, but when you're faced with starvation, they're better than nothing. If you're planning on cooking in your shelter, you'll need to vent the kitchen area to insure dangerous gases don't build up. As a side note, some heaters and generators give off carbon monoxide. So, make sure any devices you put in your shelter are safe.

This brings up the final issue: breathable air. As with water, if you're going to have to remain underground for months (maybe a year), it is highly unlikely you'll be able to have air tanks big enough to keep you and your family alive. Also, each time you breathe out, you boost the level of carbon dioxide in the air. If that level gets too high, it can be as lethal as carbon monoxide. The solution is an air purification system. It can be connected to the outside to both draw in fresh air (and filter it) and vent excess carbon dioxide. There are filter systems that are military grade - meaning they can remove NBC (nuclear, biological and chemical) contaminants.

Finally, as with a standard safe house, you're going to need to stock your shelter will all of the usual items:

- Clothes
- Dishes
- Medical supplies
- Lights (glow sticks, lanterns, candles etc)
- Tools and supplies
- Bedding
- Personal items (bank books, deeds, stocks, bonds, ID, passports social security cards etc)
- Books, games, toys, and other items to pass the time
- The key element that will be different for an underground shelter is that you'll have to conserve space as much as possible.

WATER PURIFICATION

As previously mentioned, water is one of the critical elements in survival. You need two quarts a day to drink, more if you're active, and you can't go without it for more than three days - or you're dead! Also, your body needs water to digest food. So, every time you eat, you should also drink something. In the event you're short on water, forgo eating as long as possible. Remember, you can survive longer without food than you can without water. Being hungry is unpleasant, being too dehydrated is deadly!

Water is also a very heavy compound; so carrying enough to last days (or even weeks) for every member of your family is virtually impossible. This means purifying water in whatever place (type of environment) you end up in following a disaster.

Fortunately, there are a whole host of methods available to make water potable (drinkable), and many of them are small, inexpensive, and easy to use. Let's take a look at just a few of the basic systems:

- Filters: these can be a large scale filtration system that uses a power source (often a gas engine) to pump brackish water through a system of filters, and give you clean water out the end. The problems with this scale of a system are its size and its need for fuel. Now, if you can afford to set up a complex system, have plenty of fuel, extra filters (and spare parts) - then go for it! On the other hand, there are many inexpensive filters on the market (a simple search of the internet using the keywords "water purification system survival" will give you many to choose from). Most will filter many gallons of water (several *thousand*, in fact) before needing to be replaced. These often work on simple gravity - just pour the water in the top, and collect the clean water in a pot placed under the filter. Remember, a pump has a *lot* of moving parts; the more parts, the more chances of something wearing out or breaking. Unless you have spare parts, tools, and the knowledge to repair the pump, go with as simple a system as possible.

- Boiling: this is the simplest and the most old-fashion of methods for purifying water. For this, simply scoop up water from a fast-moving river or stream (the faster the better), as this will help to carry away any pathogens that might be growing in the water. You want the water to *look* as clean as possible to start with, and then bring it to a rapid boil for 20 to 30 minutes. This will kill just about all water-borne diseases. Now, it will not remove contaminants like heavy metals, radiation or chemical agents. However, if you know the water source is fairly clean, then this system can work.

- Chemicals: there are plenty of pills sold in camping stores that can purify water, but they generally have a limited shelf life (especially once you open the bottle). Still, for a short term situation, they can work.

- Home-made System: in a pinch, you can cobble together a filter from your own gear. Take a pot, punch some small holes in the bottom, and then place a cloth in the bottom. Put a couple inches of sand or loose dirt on top of that, and then more cloth. Add small stones on top of the upper cloth, and then put a clean bucket under the device. After that, pour water in the filter system and collect it in the lower bucket. Next, take some household bleach (not one with any special chemicals), and add 8 drops for each gallon of water. This will further purify the water. While not perfect, this system is better than nothing, and can be quite effective when used on water that is fairly clean to start with.

As a final note, drinking large amounts of seawater is deadly! The high salt content will heighten your level of dehydration. However, small amounts of seawater while you're completely healthy are not dangerous. A single eight-ounce glass of it each day isn't harmful, and can restore vital salts and minerals to your body.

FIRE CRAFT

As previously mentioned, fire is of critical importance for survival. You cook with it, it keeps you warm at night, and it can protect you from danger. With any type of fire creator, there are always three factors to consider:

- Fuel
- Heat
- Air

You have to have all of these at all times, or the fire won't start and/or will go out. No matter what you use to start your fire, there are some common elements to consider. So, let's examine them:

- What's the terrain? If you're in a forest with lots of leaves, twigs etc lying around, clear the area you want to build your fire in.
- Security: keep the fire close to your shelter, but not too close that the fire can spread to it.
- Protect the fire from the wind.
- Position the fire so it delivers heat in the direction you need.
- Have a supply of fuel nearby.

There are three kinds of fuel, and you'll need all of them to start and maintain the fire:

1. Tinder: this is fast burning material like straw, sawdust, dead pine needles, and so on.
2. Kindling: this material burns easily, but lasts longer than tinder, and will help to raise the temperature of your fire hot enough to ignite the fuel. This is material like small twigs, split wood, heavy cardboard, and wood doused with oil, wax or some other inflammable liquid.
3. Fuel: this is the material that will burn longest and give you the best heat. You can use dried branches, spilt logs, long dry grasses that have been tied into knots, dried dung, fats, and coal.

If you have time, build a fire wall with wood or stones to direct the heat in the direction you want. Also, it will shield the fire from being seen by the outside world.

In terms of building the fire, you want to put down a base of tinder, surround that with your kindling, and then put your fuel on top/around that. Light the tinder, and then feed the fire as needed to keep it burning until the fuel catches fire.

To actually start the fire, there are several methods you can use. Let's look at each of them:

- Modern Methods: this includes any kind of lighter, and they are pretty basic - you activate them, they ignite.
- Matches: again, pretty basic; just be sure you have waterproof ones, and you want the "strike anywhere" kind.
- Convex Lens: take a magnifying glass or the lens out of a camera, telescope, binoculars etc and concentrate the sun's rays on your tinder. Focus the rays on the same spot until it starts to smolder, and then gently blow on it until it catches fire.
- Metal Match: this is a metal device that you use with a knife to create sparks. Place the match in the tinder, and scrape the knife against it. As sparks fly into the tinder, blow on it until it catches fire.
- Battery: this works best with a car battery. Attach wires to each terminal, and then strike the ends together in the tinder to make sparks. Be sure the wires are insulated, or you're wearing gloves. If you have some steel wool, attach the ends of the wires to a piece of it, and stick it in among some tinder. The steel wool will heat, and eventually start a fire.
- Gunpowder: this can be dangerous, if you have to extract the powder from a live bullet; so, use this method sparingly. Spread the powder in the tinder, make some sparks, and it will catch fire.
- Flint and Steel: this is similar to the metal match process. Strike the flint and steel against each other close to the tinder until the sparks make the tinder smolder. Blow on it gently until the fire starts.
- A Fire-Plow: this is among the most primitive method of starting a fire, but it can be effective. Take a piece of

soft wood and cut a shallow groove down the middle. Make a blunt tip on a thin piece of hardwood, and then rub it up and down the groove as fast as possible. As the wood heats up, sprinkle some tinder across the soft wood, and blow on it.

- Bow and Drill: this requires a bit of construction work to build, but it is very effective. Take a stick about two feet long and half an inch thick, and tie a cord to each end. Make it short enough that the stick is bent into the shape of a small bow. Next, for the drill, get a piece of hardwood that's about the same size as the bow, make one end of it blunt (to make good friction), and then wrap the bow cord around it. Finally, get a small stone with a slight depression in it to act as the socket. To start a fire, put the blunt end of the drill against a piece of wood, press down on the other end of the drill with the socket, and then pull the bow back and forth. This will spin the drill into the soft wood, and eventually - as you go faster and faster - heat will build up. Have some tinder around the drill, and it will catch fire once enough heat has built up.

FOOD PROCUREMENT

When hunting, focus on small animals - they're easier to catch, there are more of them, and they're easier to clean and cook. For large animals - deer, elk etc - only hunt those if you have experience and the proper weapons.

Nearly all small animals are edible; few are in any way poisonous. A key to catching them is knowing their habits and patterns of behavior. Most live in a nest or den, in a certain area, and feed and drink in fixed places. Once you learn where these places are, it'll be easier to catch them.

Just about any animals that walk, crawl, swim or fly can be eaten. Your issue will be overcoming the natural aversion you have to eating something that looks (for lack of a better word) *icky*. However, *hunger* is a great motivator; so let's look at what you *can* eat (not what you *prefer* to eat):

Insects
They are the most abundant form of life on Earth, and they are easy to catch. Their bodies are 65% to 80% protein (beef is only 20%). So, while not appealing, insects can be an important food source. However, avoid adult insects that can bite or sting, brightly colored or hairy insects, and insects and caterpillars with a pungent odor. Insects with bright colors are often toxic, as are those with a foul odor and that are hairy. Also, don't eat spiders, and avoid flies, ticks and mosquitoes; all of these often carry diseases.

The hard shells of many insects often carry parasites, so remove the wings and legs (if they have barbs), and cook the insects. If eating insects is truly distasteful to you, you can grind them up and blend them with vegetables.

Another aspect of insects that you can enjoy is honey. You can either gather it from natural hives and/or set up hives on your own. Not only is honey a good natural sweetener, but it can also help you deal with hay fever. So long as the bees make the honey from one of the plants you (or a family member) are allergic to, the honey acts to protect you from the symptoms.

Worms
While even more unappetizing than insects, worms are a great source of protein. They can be found in damp soil, and they also come out on the ground following a rainstorm. You can easily clean them by dropping them in clean water for a couple minutes. While worms can be eaten raw, few people are able to stomach them in that form. You can dry them out, grind them up, and then mix them with other foods to boost your protein.

Crustaceans
Even if you live in a mountainous area, there can be freshwater shrimp to catch. There are also crayfish, which are similar to lobsters. If you move to an area near the ocean, there are lobsters, crabs, and shrimp. Typically, you find them close to the shoreline, in water no deeper than 10 meters. Shrimp are best caught by scooping them up in a net. Lobsters and crabs can be caught in a simple trap, which you can possibly scrounge from a deserted dock. Both crabs and lobsters are nocturnal, so your best bet at catching them is at night.

Mollusks
These are animals like the octopus, snails, barnacles, clams, mussels, periwinkles, and sea urchins. River snails are found in streams, rivers, and lakes; while in the sea, tidal pools and rocks along the beach often have shellfish clinging to them. Now, one word of warning regarding mussels: during the summer, they can be poisonous! So, avoid them during this time. In general, you can boil, steam and bake shellfish in their shells, and they can be used to make stews. However, as you may not know the conditions the mollusks have lived under, do not eat them raw or take them from water that looks polluted.

Fish
These are yet another good source of needed fat and protein. To be a success at catching them, it's best that you learn the habits of the fish. They usually feed before a storm but they

don't like to feed when the water has become muddy. Some fish are attracted by light at night. In a heavy current, they'll find an eddy to rest in, which is usually near rocks. They also like to gather in deep pools, under brush overhanging the water, and around submerged objects (foliage, logs etc) that give them shelter.

No freshwater fish are poisonous, but they can have parasites. So, cook them thoroughly. The same goes for ocean fish that live in a reef or are near a source of fresh water, such as a river that empties into an ocean. If you have a boat and are able to catch fish that live farther out in the ocean, they won't have parasites, and you can eat them raw.

However, some types of saltwater fish are poisonous, and need to be avoided. Don't eat the porcupine fish, cowfish, thorn fish, oilfish, jack, and the puffer. The barracuda isn't poisonous, but it can give you ciguatera (fish poisoning) if you eat it raw.

Amphibians
You can find frogs and salamanders around any fresh water body. Frogs tend to stay close to the water's edge, and some are poisonous species. Don't eat brightly colored frogs or ones with a distinctive "X" marking on their back. Now, frogs are quite different from toads, which normally live in a dry environment. Several types of toads secrete poison through their skin; so don't eat or even handle toads.

Salamanders come out at night, and you can catch them with a light. They can be as small as a few centimeters to more than 60 centimeters. You generally find them in the water, near mud banks and rocks.

Reptiles
These are a fine source of protein and not relatively hard to catch. They should be cooked, to ensure no parasites on their skin or in their flesh infect you, but if starvation is a factor, they can be eaten raw.

One reptile you should definitely not eat is the box turtle. Its diet consists of poisonous mushrooms, and it can have a high level of poison in its body. Don't bother trying to cook it, that doesn't destroy the poison. Also don't eat the hawksbill turtle, common to the Atlantic Ocean; it has poison in its thorax gland. There are also various snakes, sea turtles, alligators, and crocodiles that contain poison. Unless you're an expert in identifying these, don't bother hunting them.

Birds
On the plus side, you can eat any species of bird you can catch. The only downside is the flavor - it'll vary quite a bit, and some might be what are known as an "acquired taste". If you catch birds that eat fish, skin them - it'll improve the taste. Some birds, like the pigeon, sleep very hard at night, and you can easily pluck them from their nests using just your hand.

During nesting season, some birds won't leave their nest, even if you walk right up to them. To find the nests, watch the directions the birds fly. They tend to go from their nest to water, to their food source etc. Not only can you figure out where to look for their nests, but if you have nets, you can put them up to catch the birds as they fly. The nests can also be a source of eggs. However, do not remove all of the eggs; the birds will abandon the nest. Leave two or three, and mark those eggs being careful not to actually touch them with your hands or the mother will abandon them. The bird will then lay more, and you can return to remove more eggs. It's important that you always leave the marked eggs; that way the birds will be able to produce young.

Mammals
These are another great source of protein, and among the tastiest. However, they have some drawbacks. Mammals are also the smartest of animals, and they are very good at detecting traps and snares you put in place. Also, their ability to inflict injury is directly proportional to their size. Mammals have teeth, and just about all of them will bite to defend themselves. Even something as small as a squirrel can scratch and/or bite, causing a serious wound, which then risks getting infected. Finally, remember that a *mother* mammal can be very aggressive defending her young.

While all mammals can be eaten, some should still be avoided. Polar bears and the bearded seal both have toxic amounts of vitamin A within their livers. The platypus has glands that contain poison. Any mammals that live by scavenging, like the opossum, can carry diseases.

HUNTING TECHNIQUES

In order to get much of the food you need from animals you're going to have to hunt. So, you can use rifles and shotguns to hunt some animals, but in some cases traps are the best method. For large sized animals - elk, deer etc - a rifle works best. If you're going after pheasant, pigeons, and other small game birds, a shotgun works best.

In the event you're not skilled with a firearm - or maybe you live in an area where you don't want people to hear any gunshots - traps are the best means of catching game. This may sound strange, but remember - you have no idea what sort of disaster is going to befall society. If you get into a situation where you need to hide yourself and your family away, hunting with a gun will not be possible. So, let's look at some types of traps, and how to set them up.

Traps
First, before you even prepare a trap, you must figure out several factors:

- Become familiar with the animals you want to catch
- Know how to build a proper trap
- Don't alert the animals to your presence by leaving any signs

There's no one trap you can put out to catch all animals. You have to figure out what species are in your area, and then set traps tailored to catch them. Here's what to look for:

- Trails and runs
- Animal tracks
- Animal droppings

- Vegetation that has been rubbed against or chewed
- Signs of roosts or nests
- Areas where animals feed and get water

Put your traps where you've seen proof of animals passing through the area. There are two different paths you can see - a run or a trail. The former is something only one species uses, and the latter will have signs of several species traveling through it. Even the best constructed trap won't catch a thing if you don't place it properly. All animals have places they sleep, places they get water, and places they feed. In every case, there are trails connecting them. By placing traps in these areas, you'll have a good chance of catching game.

First, the trap needs to be concealed, but you have to be careful not to disturb the area; this will alert the animals and they'll avoid the trap. Build the trap away from where you're going to set it; then carry it there and set it up. Use old, dried out wood/ vegetation to make the snare. Sap will bleed from newly cut vegetation and give off an odor that animals will smell.

You must also mask your scent on the trap and in the area around it. Almost all mammals you'll want to hunt use smell more than sight. So, even a slight human smell will warn them of your presence. The urine from a previous kill will work best, but don't use human urine! If that's not available, get mud from a place with lots of rotting vegetation. Cover your hands with the mud as you place the trap, and also coat the trap. If you can't do either method, set the trap outside to weather for a couple days before using it. Be careful not to touch the trap while it's weathering.

When you place the trap on the trail, you'll want to build a channel. Make funnel-shaped walls from both sides of the trail to the trap, the narrow part of the channel should be closest to the trap. Most wild animals won't back up; they prefer to face forward. The channel doesn't have to be impossible for the animal to get through, merely difficult.

Use of Bait
While you can set a trap without bait, you've got a better chance of catching something by using bait. It needs to be some food item the animal knows. However, don't make

it something the animal can easily find in the area. As an example, putting corn in a trap that's in a corn field probably won't work. On the other hand, using corn if it doesn't grow in the area at all probably won't work either. An ideal bait is the peanut butter found in a MRE. It's also a good idea to spread some of the bait outside the trap. The animal will sample it, and then start to crave it. This will help the prey overcome its fear of the trap.

Trap Construction
Your basic trap is designed to choke, crush, entangle or hang an animal. Ideally, you want to incorporate at least two of these methods in each trap. You also want the power mechanism to be very simple. Remember this: the more complex something is, the more prone it is to breakdown. So, if the trap uses the struggles of the animal, gravity, or the tension in a bent sapling as power - that's perfect.

The most important element to the trap is the trigger. When you're preparing a trap, there are three questions to ask yourself:

- How will it affect the prey
- What power source are you going to use
- What sort of trigger will you use

Simple Snare
The basic snare has a noose that you place across a trail or over the opening to an animal's den, and then attach it to a stake firmly planted in the ground. If you use some sort of cord to make the noose, small twigs and/or blades of grass should be used to hold the noose open. If you can get hold of a spider's web, the filaments are ideal to use for that. The noose needs to be big enough to slip over the animal's head. When the animal struggles, the noose starts to tighten, and then you can capture the animal when you come back. In general, unless you use wire, the snare probably won't kill the animal. For that, you're going to need a killing device.

Rabbit Stick
This is a simple and effective killing device; just a stout stick about as long as your arm.

Spear

This is essentially just a rabbit stick with a sharp point - either one tip sharpened or with some sort of sharpened head secured to the end. Unless you truly become proficient at throwing the spear, only use it as a jabbing and thrusting implement.

Bow and Arrow

This is a great weapon to have on hand. Not only can you use it to kill small game, but it can also be a means of personal protection. It is silent, effective, and deadly. Also, if you're going to be out in the wilderness for an unknown period of time (you have no way of knowing how long the disaster will affect society), the bow and arrow is better than firearms for getting game. At some point, guns run out of ammunition, no matter how many you buy, and they need to be cleaned and maintained. With a good bow, you can keep using it as long as you retrieve the arrows or make more.

Go to a decent sporting goods store and buy several bows and a good supply of arrows. After that, if you have time, set up some targets in your yard and learn how to shoot. If you don't have time - you buy the supplies shortly before leaving for your safe house - set up a target out in the wilderness and practice!

If you have to travel to safety at a moment's notice and are unable to buy any bows and supplies, you can make them fairly easily. Find a piece of dead, dry hardwood about three feet long, and that doesn't have knots or limbs. Scrape the thick end until its pull is the same as the thin end. Use a tough cord with good flexibility as the bow, and attach it to the tips of the bow. You want the tips to end up bent inward a good amount, but not too much - otherwise the bow will not have much strength to it.

For arrows, use dry sticks about half the length of the bow, as straight as possible, and scrape them smooth. You can make them straighter by slightly heating them over a fire or hot coals. Hold the arrows as straight as you can and let them cool. Notch one end of each arrow so they can be held by the bowstring, but do *not* split the wood! Either cut or use a file to make the notch. If you want, try to add feathers to the arrows, as they will improve the accuracy of the arrow's flight. However, feathers are not vital. As for the arrowheads, if you can, buy some ahead of time. If not, things like sharpened rocks, glass, bone, and metal can be used.

Sling

This is another simple killing device. Take two long pieces of cord, and tie each to opposite sides of a piece of cloth, leather or canvas. The cloth should be about the size of your palm. By placing a rock in the cloth, you can then swing it around and around by holding the two cords. When you're ready to throw, let go of one of the cords. Over time, you can become very proficient and accurate in your throws. A sling is a very good means of stunning and killing small game.

FISHING TECHNIQUES

Ideally, you want to evacuate to your safe house with a full set of fishing gear: poles, line, reels, lures, nets etc. But, if that isn't possible, you need to be prepared to cobble together your own gear.

Fishhooks
A basic hook can be made from needles, pins, wires, small nails, and any bit of metal. For that matter, things like bone, wood, coconut shells, shells, thorns, and even turtle shell will do.

A very basic, easy-to-make type is a bit different from the hooks you see in the store. Simply take a small length of material (bone, metal, wood etc) and sharpen both ends. Then tie your fishing line to the middle of the hook and place bait on it. This is what is known as a gorge. You need to be sure you make its length short enough that the fish can swallow it. When it does, the points stab into its throat, and you can reel it in.

Stakeout
This is a method of fishing that will allow you to fish without spending time at the activity. Place two supple branches in a lake, stream or pond, making sure that they're well set in the bottom. Next, tie a cord to each branch, positioned so it is just below the surface of the water. Then tie a number of short lines to this rope, and put gorges or hooks on their ends. Be sure to spread them out enough so they don't get tangled. Put bait on the hooks and/or gorges, and leave for several hours.

Fish Traps
You can either buy some of these to put in your safe house, or make one. A fish basket is made by tying several sticks together using rope or vines, and forming it into a funnel. Close the wider opening, making sure the small hole is big enough for fish to get through. Set the trap in the body of water (if it's a

fast-moving stream, be sure to anchor it), place some bait in it, and then check on it daily.

Spearfishing
This only works in shallow water that's roughly waist deep, and the fish you want are good-sized and plentiful. If you can't buy a decent spear, you can make one out of a straight, long sapling. You can either attach a blade (a knife, a sharp piece of metal etc) to one end, or sharpen the end to a point.
To use the spear, look for an area with fish gathered. Put the spear tip just below the surface of the water and then slowly edge toward a fish. Give a quick thrust and impale the fish, pushing it down to the bottom of the stream. Don't try to bring the fish to the surface with the spear; it'll probably slide off. Holding the spear with one hand, grab the fish and hold it with your other hand. Also, don't try throwing the spear, this is especially important if a knife is the point of the spear. You might lose or damage the blade. Finally, remember that light refracts through water; so an object you're looking at will appear to be in a slightly different place than it really is. With practice, you'll find you become very proficient.

Chop Fishing
This is a simple, easy way to catch fish that are in a large school. At night, wade into an area full of fish. Fish are attracted to light; so use one to bring them closer to you. Use a machete or some other type of large blade weapon as essentially a club. Smack the fish with the side of the blade; this will stun them. If you use the blade, you'll end up slicing them into pieces, and lose the fish in the dark.

Fish Poison
This is a simple, easy way to gather fish. You can buy fish poison or make your own. Some plants have rotenone in them. This substance stuns or kills fish (and other cold-blooded animals), but it won't harm you when you eat it. Use rotenone in a pond or small stream - at its headwaters. One word of warning, the poison doesn't work if the water is below 50 degrees F. Here are plants you can use:

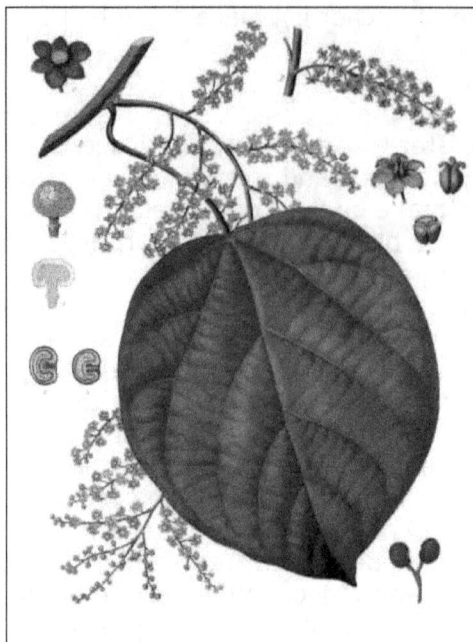

Figure 17. Anamirta cocculus.

<u>Anamirta cocculus</u>: is a vine that grows in southern Asia and islands of the South Pacific. Take the bean-shaped seeds, crush them, and toss them in the water.

Figure 18. Croton Tiglium.

<u>Croton tiglium</u>: a shrub/ small tree that grows on islands of the South Pacific. Crush the seeds and toss them in the water.

Figure 19. Barringtonia

<u>Barringtonia</u>: this is a large tree that grows near the ocean in Malaya and parts of Polynesia. Take the seeds and the bark of the plant, crush, and toss in the water.

Figure 20. Derris eliptica.

<u>Derris eliptica</u>: this is a tropical shrub and vine. Grind up the roots to make a powder, and then mix with some water. Toss a large amount into the water.

Figure 21. Duboisia.

<u>Duboisia</u>: a shrub, it grows in Australia, and it has white clusters of flowers; the fruit is berry-like. Grind up the entire plant, and toss into the water.

Figure 22. Tephrosia.

<u>Tephrosia</u>: this small shrub has bean-like pods, and it grows in the tropics. Take the leaves and stems, crush or at least mangle them up, and toss in the water.

<u>Lime</u>: this is one of the easiest to use, and to get ahold of. Lime is available in many stores, or you can make your own by burning seashells or coral. Toss it in the water and wait for the fish to float to the surface.

Figure 23. Black Walnuts.

<u>Black Walnuts</u>: take the husks, crush them up, and then toss them in the water.

POISONOUS PLANTS

Once you're out in your safe house, the last thing you're going to want happen is getting sick, or even just getting a rash from a plant. Staying healthy will be critical to your survival. So, you need to know what plants are dangerous and to be avoided. There are three ways that plants can affect you:

- If you eat them
- If you inhale their spores or other vapors
- If they touch your bare skin

Here is a listing of the plants to avoid:

Figure 24. Castor bean.

<u>Castor bean, aka the castor-oil plant</u>: a semi-woody plant with star-like leaves. It has fruit at its top that grow in clusters. Every part of the plant is poisonous to eat.

Figure 25. Chinaberry.

<u>Chinaberry</u>: a tree that grows tall with compound leaves and flowers that have a dark center surrounded by light purple. It produces orange colored fruit that are about the size of marbles. Every part of the tree is dangerous, if you eat them. However, the leaves can be used as an insecticide. Place some of the leaves in your food storage area, just be sure not to eat them.

Figure 26. Cowhage.

<u>Cowhage, Cowage, aka Cowitch</u>: a vine-like plant with oval leaves that are in bunches of threes. It also has dull purple flowers and spikes, and the seeds come in hairy pods. The pods and flowers will irritate your skin, and can cause blindness if you get them in your eyes.

Figure 27. Death camas or death lily.

<u>Death Camas, Death Lily</u>: an onion-like plant, its leaves are grass-like. The flowers have six-parts and green petals with a heart-shaped structure. Every part of the plant is extremely poisonous.

Figure 28. Lantana.

<u>Lantana</u>: a shrub-like plant with round leaves and flowers that grow in a cluster. The flowers can be red, pink, orange, yellow or white. The fruit is black or dark blue, and berry-like. The plant has a very strong scent. If you eat any part of this plant, death can result; it is extremely poisonous.

Figure 29. Manchineel.

<u>Manchineel</u>: is a tall tree (over forty feet), and it has green shiny leaves and small flowers that are greenish in color. The fruit is green-yellow or green. The entire tree is very toxic, and causes skin irritation within half an hour of exposure. If you burn the wood, the smoke will irritate your eyes, and even moisture drops from the plant can cause skin inflammation.

Figure 30. Oleander.

<u>Oleander</u>: comes in a small tree or shrub variety with dark, very straight green leaves. The flowers can be red pink, white, or intermediate colors. The fruit is brown, pod-like, and often have lots of tiny seeds. Every part of the plant is extremely poisonous, even the wood should not be used to cook. The smoke from the wood can poison your food.

Figure 31. Pangium.

<u>Pangium</u>: a tree with heart-shaped leaves that grow in a spiral. The flowers are green and grow in spikes, and the fruit is brownish, large, and pear-shaped. The fruit also grows in clusters. All parts - especially the fruit - are poisonous.

Figure 32. Physic nut.

<u>Physic nut</u>: a small tree or shrub, it has alternate leaves that come in three to five parts. The flowers are small and greenish-yellow, and the apple-sized fruit is yellow and come with three big seeds. While the seeds have a sweet taste, their oil is a powerful laxative, and all parts of the plant are a poison.

Figure 33. Poison hemlock or fool's parsley.

<u>Poison hemlock, fool's parsley</u>: an herb, the stems are smooth and hollow, and can be red, purple or mottled. The white flowers grow in groups and are small, and they form flat umbrella-like shapes. Easy to mistake for Queen Anne's lace or a wild carrot, it is extremely poisonous; a tiny amount ingested can lead to death.

Figure 34. Poison ivy and poison oak.

<u>Poison ivy and poison oak</u>: similar in appearance, they both have compound alternate leaves that have three leaflets. The poison ivy leaves are smooth; the oak's resemble oak leaves. The ivy is a vine; the oak is a bush-like plant. While not deadly, contact with either plant can cause severe skin irritation.

Figure 35. Poison sumac.

Poison sumac: a tall shrub, it has compound, alternate leaf stalks that have seven to thirteen leaflets. The flowers are ordinary and greenish-yellow, and the berries are pale yellow or white. Contact with any part of the plant can irritate the skin.

Figure 36. Renghas tree.

Renghas tree: this species encompasses around 48 different shrubs and trees, and all have flowers like poison ivy. The leaves alternate in axillaries or terminal panicles. Contact with the plant can irritate the skin.

Figure 37. Rosary pea or crab's eyes.

<u>Rosary pea or crab's eyes</u>: this is a vine with compound, alternate leaves, red and black seeds, and light purple flowers. This is a very deadly plant; a single seed can have enough poison to kill any adult.

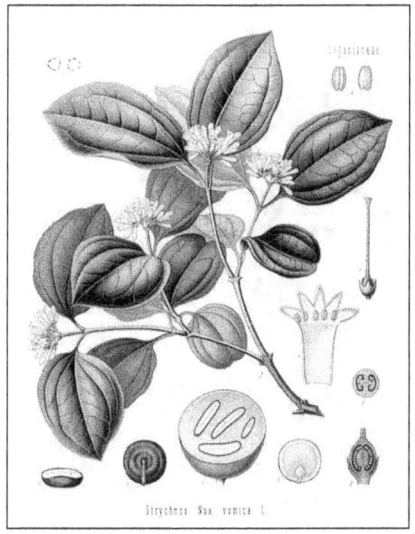

Figure 38. Strychnine tree.

<u>Strychnine tree</u>: a medium-sized tree of the evergreen variety; it has a thick trunk that is often crooked. It has deeply veined leaves that are oval and grow in alternate pairs. Loose clusters of small greenish flowers grow at the ends of its branches, and then berries that are fleshy and orange-red. Every part of the plant is poisonous, and the seeds in the berries contain strychnine.

Figure 39. Trumpet vine or trumpet creeper.

Trumpet vine or trumpet creeper: a woody vine with pea-like fruit, the leaves come in 7 to 11 toothed leaves in each stock. Trumpet-shaped flowers are scarlet or orange in color. Contact with the plant will irritate the skin.

Figure 40. Water hemlock or spotted cowbane.

Water hemlock or spotted cowbane: This herb has hollow stems that are sectioned off like bamboo. It can be red, purple or mottled. The flowers are white, small, and often form flat umbrella-shaped groups. The roots often have hollow chambers and can produce small amounts of yellow oil when cut. Another very deadly plant, even a tiny amount can lead to death.

DANGEROUS ANIMALS

As with plants, there are a whole host of animals - of all shapes and sizes - that can be dangerous to you. It's important that you know how to spot them, and keep in mind that, in the case of dangerous animals, size most definitely does *not* matter! So, let's look at each group.

Scorpions
They live in jungles, deserts, and forests around the world. Dying from a scorpion sting is rare, but children, the elderly, and the sick can succumb to them.

Spiders
The Brown Recluse can be recognized by a violin-shaped light spot on its back. The bite normally isn't fatal, but it can cause the tissue around the bite to die, and even result in amputation being needed, especially if the bite is on a finger or toe.

There are several members of the widow species around the world. The Black Widow, found in North America, is the best known. It's a small and dark spider, and usually has an hourglass-shaped marking that is red, white, or orange on its abdomen. Not usually lethal, a bite from a widow can cause pain, sweating, weakness and shivering, and being disabled for up to a week.

The Funnelwebs are big, gray or brown spiders. They usually hunt at night. The bite symptoms are similar to a widow's: severe pain, weakness, sweating and shivering, and disabling periods lasting as long as a week.

Tarantulas are hairy, large, and well known because many are sold in pet stores. Some of the South American varieties have dangerous toxins, but most only result in a painful bite. They have large fangs, which they use to capture food, but a bite from a tarantula is generally only painful. Bleeding can result, and infection can set in.

Centipedes and Millipedes
These are generally small and harmless, but some desert and tropical species can be as long as 25 centimeters. While some species have a poisonous bite, infection is the main danger. They have sharp claws that dig into the skin and puncture it when they bite.

Bees, Wasps, and Hornets
These are the most familiar stinging insects around, and they're found just about everywhere. There are many varieties, and you can easily recognize them. Other than the famed "killer bees", most are rather docile, and will not sting you so long as you leave them alone. The one plus to bees is that each can only sting once, and then they die - the act of stinging rips their stinger and poison sac from their bodies, and they die soon after. On the other hand hornets, wasps, and yellow jackets are able to sting multiple times.

If you simply avoid these insects, you can easily protect yourself. Most people will have a relatively mild reaction to the stings, and be fine in a couple hours. If you or any of your family is allergic to their venom, you could go into anaphylactic shock, maybe lapse into a coma, and you could even die. Keep some sort of antihistamine medication on hand at all times.

Fire Ants / Army Ants
Most ants are harmless. However, fire ants - as their name implies, can bite and sting you. Their venom is similar to that of bees, and if you're allergic you can have the same reaction to their sting. In the case of army ants, not only do they have a terrible sting, but they attack in massive swarms, and it is not unheard of for them to kill large animals (even livestock) and humans. Avoid both at all costs. If you get word or see the signs of a swarm of army ants coming your way, you should either evacuate or take steps to protect yourself and your family. A body of water will *not* stop the ants; they're capable of creating rafts and floating across! Instead, you can use fire to drive them off, or fill buckets with kerosene, gasoline, turpentine or other chemicals, and then set the legs of a bed

or bench in them. Be sure that there are no tree limbs above where you set the buckets and bed up (the ants will drop on you from above), and stay on it until the ants move on.

Ticks

These are round small arachnids that feed on blood. They can spread ailments like Lyme disease, encephalitis, Rocky Mountain spotted fever etc that can disable or kill. There aren't many treatments for these diseases; the best course of action is to prevent getting them by preventing ticks from feeding on you and your family. According to the best authorities, it takes about 6 hours for a tick to transmit any disease to its host. So, as long as you inspect your body each morning and evening (at a minimum), you should be fine. Ticks are often found in tall grasses and thick shrubs. So, if you go out in a field or move through dense brush, check yourself upon getting clear or getting home.

Leeches

These are blood-sucking animals that look like worms. They're found in bogs, swamps, and places thick in tropical vegetation. Also, if you catch and kill various animals (turtles and other creatures found in fresh water), leeches can be found on their bodies. When moving through a swampy area, tuck your trousers into your boots, and check yourself as often as possible. Do not eat leeches under any circumstances! This means treating water you get from a questionable source by boiling it or using some sort of chemicals.

Snakes

There's no one perfect way of identifying poisonous snakes. The reason for this is that you have to get *very* close to the snake to do so! So, when in doubt, just leave any snakes you encounter alone. Here are some safety rules to follow when you're in an area where poisonous snakes are present:

- Be careful where you walk, and watch each spot where you put your feet.
- Step on a log instead of over it.
- Don't pick fruit or move around a body of water without looking closely.
- Never molest, tease, or harass any snake. Snakes are unable to close their eyes; so you can't tell if they're

sleeping. Snakes like cobras, mambas, and bushmasters will attack if you corner them or if they're guarding their nest.

- Use a stick to turn over rocks and logs.
- Wear good solid footwear, especially at night.
- Check your bed, home, and clothing carefully.
- Snakes can't hear, which means you can easily surprise them. So, if you encounter one, stay calm, and slowly back away from it. Most snakes will move away from you, if you give them the opportunity.
- If you are desperate for food and must hunt snakes, be very careful while doing so. Use a forked stake or stick to pin their head down, and then kill them.

In terms of locations, there are a number of poisonous snakes around the world. Here is a listing of them, based on the continent you live on:

The Americas
- American Copperhead
- Bushmaster
- Coral snake
- Cottonmouth
- Fer-de-lance
- Rattlesnake

Europe
- Common adder
- Pallas' viper

Africa and Asia
- Boomslang
- Cobra
- Gaboon viper
- Green tree pit viper
- Habu pit viper
- Krait
- Malayan pit viper
- Mamba
- Puff adder
- Rhinoceros viper
- Russell's viper
- Sand viper

- Saw-scaled viper
- Wagler's pit viper

Australia
- Death adder
- Taipan
- Tiger snake
- Yellow-bellied sea snake

Dangers in Bays & Estuaries
If you're in an area where rivers empty out into an ocean or sea, you need to be wary of the dangers associated with them. In the shallows of an ocean are sea urchins, which have spike-covered bodies. If you step on one, a painful wound will result, and it can become infected. Always wear some kind of footwear, and only slide your feet along.

Stingrays can be very hazardous in shallow water. There are a wide variety of species, but all of them have a sharp spike located in their tail. In some species, it has venom, and even the non-venomous ones can give you a very painful wound if you step on one.

Saltwater Creatures
In the event your safe house is near an ocean, or you live on an island - or the global disaster is a flood - you may find yourself living near or traveling across an ocean. So, you need to know what ocean creatures to be wary of.

Sharks
While among the most feared creatures in the sea, most sharks are quite harmless. Even those that eat meat (and are considered man-eaters), can be avoided with ease. The dangerous sharks are those with a wide mouth and visible teeth. These sharks are attracted by the following:

- Blood in the water
- A fish in distress
- Thrashing/splashing about
- A person on a surfboard

Sharks can smell even a single drop of blood from literally miles away! A fish in distress tends to thrash about, but so do people swimming, which is why you need to avoid doing that if sharks are in the area. The reason for the final item is because a person on a surfboard resembles one of the favorite meals of a shark: a seal. So, avoid putting yourself in danger by forgoing surfing.

While any shark can bite and inflict a painful wound, and they can even cause a fatal injury, these are very rare. In fact, more people die from beestings each year than from shark attacks. Still, when you're in a survival situation, you need to do all you can to prevent injuries; so don't take chances around any sharks.

Tang
Also known as the surgeonfish, they have scalpel-like spines in their tails. Avoid getting close to them, as any wound they inflict can become infected.

Toadfish
These are dully colored fish that bury themselves just below the surface of the sand and wait for prey to come by. As they have toxic and sharp spines running down their backs, the danger with them is stepping on them. Serious infection and even death can result.

Stonefish
This fish is much like the Toadfish in that it tends to lie on the ocean bottom, it has venom in its spines, and (as its name implies) it looks like a stone.

Jellyfish
Death from jellyfish stings is rare, but it does happen, and at the very least, they'll inflict an extremely painful sting. One of the largest is the Portuguese man-of-war. It looks like a large purple or pink balloon just floating on the water. A typical specimen will have poisonous tentacles more than 30 feet long hanging below its body. Stay away from its tentacles, and those of any other jellyfish and you'll remain safe. Also, do not touch the tentacles of a jellyfish that's washed up on shore, even if it appears to be dead; they can still sting!

Barracuda
These are far more dangerous than sharks; as they are prone to attack people for no reason. Also, avoid eating them because they sometimes carry a poison called ciguatera, which is deadly.

BASIC SURVIVAL MEDICINE

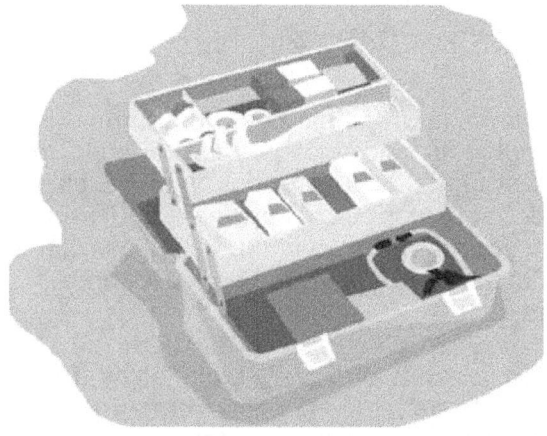

Ideally, you'll want to have at least a couple first aid kits with you at your safe house. However, if evacuation is done on short notice, or you're out at the safe house for a very long time, you may either not have a medical kit or you may run out of some supplies. So, with that in mind, here are some basic means of caring for you and your family.

In the event of choking, you need to open the airway and maintain it. To do so, use the following steps:

1. Check to see if the victim's airway has any sort of obstruction. If the person can speak or cough, they can clear the obstruction on their own. Often times, raising one arm above their head will help to open their airway. Stand next to the person, reassure them, and be ready to help them, if necessary.
2. If their airway is obstructed, and the person has passed out, give them several thrusts to their abdomen until the obstruction clears.
3. Use a finger to sweep their mouth to remove foreign objects, dentures, broken teeth etc.
4. Grasp the victim's lower jaw, one hand on each side, lift, and move the jaw forward.
5. Once their airway is open, pinch their nose closed using your thumb and forefinger, and then blow two breaths in their mouth. Let their lungs deflate following the second blow, turn your head, and do the following:
   - *Look* to see if their chest rises and falls.
   - *Listen* for air coming out as he/she exhales.
   - *Feel* for airflow on your cheek.
6. If the two breaths don't cause spontaneous breathing,

continue mouth-to-mouth resuscitation.

7. The person may vomit while you're performing mouth-to-mouth resuscitation. Check their mouth from time to time, and clear as necessary.

Controlling Bleeding

Out in the wilderness, or in an urban survival situation, controlling serious bleeding as quickly as possible is vital. Under normal conditions, replacing fluids is easy, but when hospitals are out of reach - or destroyed or out of supplies - and you have only yourself and your family to rely on, blood loss can lead to death - sometimes within minutes. There are three types of external bleeding:

1. *Arterial:* the blood vessels that carry blood away from the heart are called arteries. The blood from an arterial cut is *bright red* and issues forth in *distinct spurts* (pulses) corresponding to the person's heartbeat. As the blood is under pressure, large amounts can be lost in a short period of time, especially if the wound to the artery is very large. As a result, an arterial bleed is the most critical type. If you don't get it under control fast, shock and even death can result.

2. *Venous:* the veins return blood to the heart. If the blood coming from the wound is a *dark red or maroon, or bluish* in color, then a vein has been damaged. As veins are the blood vessels furthest from the heart, the flow from the wound will be slow, and much easier to control.

3. *Capillary:* the capillaries are very small blood vessels connecting the arteries to the veins. This type of bleeding is usually found in minor scrapes and cuts. A simple adhesive bandage can control this type of wound.

To control external bleeding, there are five methods you can employ:

- Direct pressure
- Indirect (pressure points) pressure
- Elevation
- Digital ligation
- Tourniquet

Now, let's look at each in detail:

Direct Pressure

This is the most effective way. You simply apply pressure directly to the wound. It must be applied long enough to seal the wound, and firmly enough to stop blood from flowing. If the bleeding goes on for 30 minutes, you need to add a pressure dressing. This is a thick amount of gauze or other material; you apply it over the wound, and hold it there by wrapping a bandage tightly around it. It needs to be tighter than a standard bandage, but don't make it so tight that it cuts off (or slows) circulation. If the person says that their limb (arm or leg) feels like it's going to sleep - the bandage is too tight. After the dressing is in place, *don't remove it;* even if the dressing gets soaked with blood, it needs to stay in place for one to two days. After that, you can take it off and put a new, smaller dressing in its place.

If you're going to be at your safe house for a long time, change the dressing daily, and check the wound for signs of infection. If the wound starts to stink, if it becomes discolored, infection is setting in, and you need to clear out the wound with an antiseptic and give the person penicillin or some other antibiotic.

Elevation

This is a simple, direct means of slowing the loss of blood, but it is only effective on a limb. Simply raise the injured arm or leg as high as you can above the person's heart. But, this won't - on its own - completely control the bleeding. You also need to apply direct pressure to the wound. If the wound is a snakebite, put the extremity below the heart; this will slow the spread of the poison.

Pressure Points

These are locations on the body where main arteries are either passing right over a prominent bone or near the surface of the skin. By pressing your finger on the pressure point nearest to the wound, you can slow arterial bleeding, and then apply a pressure dressing. If you don't know the locations of pressure points on the body, or don't have a medical book to guide you, follow this rule: put pressure on the end of the joint right above where the wound is located. These are places like: the wrist, an ankle, and the neck. However, here is a critical word of warning: do *not* apply pressure to the neck for too long; doing so runs the risk of causing the person to pass out or even die! You can maintain pressure on one of these points by putting a small stick at the joint, and then bend the joint around the stick. Hold the stick tightly in place by tying the joint closed.

Digital Ligation
You can use your fingers to stop significant bleeding by applying pressure on the vein or artery. Keep the pressure up until the blood ceases to flow, or at least slows down to the point where you can elevate the wound or apply a pressure bandage.

Tourniquet
You use one of these only when the other methods don't control the bleeding. The danger is that if you keep a tourniquet in place for too long, you'll kill the damaged tissue, which can lead to gangrene. Once that happens, amputation may be needed, and you don't want to have to do that! If you put the tourniquet on incorrectly, you can permanently damage the nerves and other bodily tissues at the point where you apply it.

Place the tourniquet around the arm or leg, between the heart and the wound. You want it a couple inches above the wound, never right over it. Use a cloth and tie it loosely around the limb; then slip a stick under the cloth to act as a handle. Turn the stick to tighten the tourniquet, but only tighten it enough to stop the flow of blood. Next, either have someone hold the stick or tie it to the arm/leg to keep it from unwinding. You can then clean the wound and put a bandage on it. Be sure to release the tourniquet every 10 to 15 minutes for about a minute or two; this will prevent loss of the limb by resuming blood flow.

Prevent and Treat Shock
Many people don't know this, but shock can kill just as easily as any wound or serious accident. Here's a little history lesson: during the Napoleonic Wars, army surgeons found that shock was killing more men than those that fell in battle. As a result, they invented the ambulance to carry men to field hospitals for quick treatment. In essence, they created the first MASH units.

Any time someone is injured, you need to assume that the person is suffering from shock - no matter what kind of symptoms they display. At its most basic, shock is simply bleeding; in this case, it is *internal* bleeding. In minor cases, it will appear as nothing more than slight discoloration to the skin. When you get a bruise, that's internal bleeding - minor, but still bleeding. In more serious cases, the bleeding is excessive; large areas of skin become discolored, and the body even distends due to the build up of blood and/or bodily fluids under the skin. Here are the first two steps to carryout when dealing with shock:

- If the person is awake, put them on a flat surface and elevate their legs about 18 inches.
- If they're unconscious, put them on their side or stomach, and be sure their head is turned to one side. This is to prevent them choking on blood, vomit, or other fluids.

If you're not sure as to the best position to put someone in, lay them out lying flat. Now, this is important: when you lay them down in the shock position, don't move them. People in shock will often shiver - as their body is losing vital heat. So, keep them warm by covering them with a blanket. If that isn't enough, add heat by building a fire or putting a space heater nearby. If they're wet, take off their clothes as soon as you can and dress them in dry clothes - if possible. In the event the person is badly injured and moving their limbs will cause pain and/or worsen their injuries, cut their wet clothes off and just wrap them in blankets. If the person is awake and able to keep down food, give them a warm drink, a hot meal, or even warm water to help them warm up. Do *not* give them an alcoholic drink! Things like brandy are supposed to warm you, but that's just a temporary sensation; alcohol actually cools the body. You can also heat small stones in a fire and put them around the person, or just lie down next to them and use your own body heat to help keep them warm.

If you're outside and unable to get to your safe house, construct a lean-to, put up a tent, or build some sort of shelter to protect them from the weather. As a side note, anytime you and/or other family members leave the safe house to go do something - tell the others where you're going and when you plan to be back. This way, if you and one other person are off doing something, and one of you is injured, you'll know that help will eventually come.

If the injured person is awake, give them small amounts of warm water with either sugar or salt in it. However, if the injury is to their abdomen - or the person isn't awake - do *not* give them any fluids through their mouth. If you or someone has the knowledge to set up an IV (intravenous) solution, use that as a means of getting vital fluids into the patient. After that, the person needs to rest for a minimum of 24 hours. With an internal bleed - unless you or someone has the surgical knowledge and equipment to operate - the only thing you can do is wait for the patient's body to heal the injury.

If you're injured and alone, lie down on a bed with a pillow or a pile of clothes under your knees. You need to keep your legs above your heart, but you do not want your knees locked in place; this will cut off the circulation in your legs. If you're outside, find a low spot in the ground to lie in, and try to make it some place protected from the weather.

Fractures
There are two kinds of fractures: compound (open) and closed. In the case of a compound fracture, a bone will protrude through the skin. So, not only is the bone damaged, but the person has an open wound. The order of treatment is to first set the fracture, and then treat the wound. With a closed fracture (as the name implies) there's no open wounds.

Someone has a fracture if - following a fall or blow to the body - they have pain, discoloration, tenderness, a deformity due to swelling, loss of the use of a limb, and/or grating. This final symptom is the sound and/or feeling associated with the broken ends of the bones rubbing together.

With a fracture, there are two main dangers: the nerves and/or blood vessels can be cut or compressed at the location of the fracture. So, move the damaged area as little as possible, and do it slowly and carefully. If the area below (further away from the chest area) the break goes numb, swells, feels cool when you touch it, or grows pale, and the person looks as if they're going into shock, then a major vein or artery may have been cut. This internal bleeding needs to be dealt with. So, treat the patient for shock, and then replace any fluids they've lost.

Once the limb is as straight as you can make it, you'll need to attach some sort of splint, brace or cast. Some first aid kits come with these - you can get very good kits that will have fiberglass or plaster casts and equipment. A key issue for any fracture or break is how long the patient has to wear the cast. Depending on the severity of the fracture, it could heal in as little as four weeks - or take as long as ten to twelve! Also, be sure that you have the equipment for *removing* the cast once the bones have healed!

In many instances, you'll have to use traction while you're putting the splint or cast in place, and sometimes maintain it while the bones knit. With small bones, like what are in the arms and lower legs, you can just use your hands to pull them into position. On the other hand, the muscles in the thighs are very strong. If the patient has broken their femur, you're going to need to build a traction device. Here are the steps in creating such a splint:

- Get two branches, both forked at one end, and about 5 cm in diameter. One needs to reach from the patient's armpit to about 20 to 30 cm past the bottom of their foot. The other needs to go from their crotch to the same point beyond their foot.
- Place padding between the splint and the leg, and lash it to the person's body.
- Notch the bottoms of the branches.
- Tie a short stick (a cross piece) between the ends of the splint; make sure it is at least 5 cm in diameter.
- Tie a wrap around the person's ankle and leave the two ends free.
- Tie the ends to the cross member.
- Put a short small stick between the cross member and the person's foot.
- Use the stick to twist the wrap tight.
- Keep twisting until the leg is pulled straight and is a bit longer than the other leg.
- Tie the stick in place to hold the traction.

Keep in mind; you'll lose a bit of traction over time as the materials weaken. So, check the splint every once in a while and tighten/repair as needed.

Dislocations
These are different from breaks in that the joints separate and the bones are shifted out of their alignment. Some of these can hurt terribly, impair nerves and/or circulation, and cause loss of

mobility below the injury. It's vital that the joint be put back in proper alignment as soon as possible.

The sure signs of a dislocated joint are:

- Pain
- Tenderness
- Swelling
- Discoloration
- Limited range of motion
- Deformation of the joint

Treatment comes in three stages:

- Reduce the relocation
- Immobilize the joint
- Rehabilitation

Reduction, which is also known as "setting" is the first step, and it refers to getting the bones into proper alignment. There are several methods available, but manual traction or using weights to set the bones are the easiest and safest. While initially painful, afterwards the patient's pain will be reduced, and normal circulation and function will resume. As you'll be under primitive conditions when you do this, you won't have access to an X-ray machine. Still, you can tell if the joint is properly aligned by its look and feel, and by looking at the matching joint.

Immobilizing the joint just means putting a splint on the dislocated joint after reducing it. There are two basic options: build a splint out of anything handy, or secure the arm or leg to the body. Here are some basic guidelines to follow:

- Secure the splint below and above the area of the dislocation
- Put pads between the splint and the limb to reduce discomfort
- After tying each strap in place, check the circulation below the joint to insure it is maintained
- As part of the joint's rehabilitation, take the splint off after one to two weeks. Increase use of the injured joint gradually until it is fully healed

Sprains

These are the simple overstretching of a ligament or tendon. Symptoms consist of swelling, pain, tenderness, and a degree of discoloration (typically black and blue). You treat them using the RICE method:

- Rest the injured area
- Ice, put it on the sprain for one day, and then apply heat
- Compression - wrap the injured area or put a splint on it to stabilize the area. If the injury is to the foot, leave your footwear on - unless it is cutting off circulation - as this will act as a good means of protecting the injured area.
- Elevation - elevate the sprain

Wounds

What would be a simple matter in the so-called "real world", an open wound in a survival situation can be life threatening. You not only have to deal with the blood loss and tissue damage, but the wound can also become infected from a myriad of sources. There can be bacteria on the item that caused the wound, bacteria on the person's skin and clothes, and/or bacteria on anything else that comes in contact with the wound.

If you take care of the wound properly, you can further reduce contamination and also promote healing. Here are the steps to follow:

1. Clean the wound, and do so as quickly after the wound is made
2. Remove or cut the clothes from around the wound
3. If the wound was caused by a bullet, spear or other long implement, check for an exit wound
4. Thoroughly clean the skin surrounding the wound
5. Rinse, do *not* scrub, the wound with a lot of clean water, and make sure the water is under pressure. While distasteful, fresh urine is a valid alternative, if you don't have any water

Unless you have the knowledge and expertise to suture or close a wound, and the proper equipment, it's best to use the "open treatment" means of dealing with one. By leaving the wound open you allow it to drain of any pus caused by an infection. So long as the wound drains, it usually won't become serious or life-threatening, despite how revolting it might smell or look.

So, step one, use a clean dressing to cover the wound, and put a bandage over the dressing to keep it in place. Be sure to change the dressing every day to look for signs of infection. In the case of a gaping wound, bring the sides together using adhesive tape. Cut the tape in the shape of a dumbbell or butterfly.

Even in the best of situations, you have to expect some degree of infection in any wound. The signs of infection are as follows:

- Pain
- Swelling of the wound and surrounding area
- Redness around the wound
- The patient running a high temperature
- Pus on the dressing or in the wound

This leads to the next procedure, treating an infected wound. First, if you have access to any sort of antibiotics, take them as directed (or in accordance with their directions). Next, put a warm, moist compress on the infected area. When the compress cools off, change it; the goal is to have a warm compress cover the wound for 30 minutes total. Do this every day, three to four times. Also, you need to drain the wound. Remove the dressing, open the wound, and gently probe it with an instrument that you've sterilized. Ideally, you want a proper medical instrument, but (in a pinch) a knife, a spoon, or any other metal device that can be sterilized will do. This is known as debridement - where you remove the pus and any dead tissue from the wound.

Next, re-dress the wound and bandage it. It's also important not to get dehydrated, so drink plenty of water. Keep this course of treatment up every day until all symptoms of the infection disappear.

Of course, it's possible that you won't have any antibiotics, either you'll have run out of them or maybe you weren't able to get any before evacuating. In that case, the wound could become severely infected and won't heal. At that point ordinary debridement won't be possible. So, consider using maggot therapy. It may sound revolting, but it is actually quite effective. Here are the steps:

- Expose the wound to ordinary flies for a day, and then cover it
- Check the wound each day for maggots
- Once you find maggots have developed, keep the wound covered, but check it every day
- Once all the dead tissue has been cleared, remove the maggots. Use sterile water to flush the area several times. If you don't have that available, you can use fresh urine (despite it being quite unappealing).

It's important you perform the final step as soon as you see that all the dead tissue is gone. If you delay, the maggots will begin to eat the healthy tissue. Two clear signs that this is occurring is an increase of pain and blood appearing in the wound.

After all this, check the wound at four hour intervals for a couple days to be sure all maggots are gone. After that, bandage it and treat as outlined above. The wound should then heal properly.

Skin Diseases
Things like rashes, boils, and fungal infections generally aren't very serious. However, during a global crisis, the last thing you need is a rash or something causing you or a loved one discomfort. Trying to hunt, fish, tend crops etc will be all the more difficult with one of these problems. So, here are the methods available to treat them.

Boils
Use a warm compress to make the boil come to a head. Use a sterile instrument: a knife, needle or wire (or something similar) to open the boil. Use soap and water (as hot as you can stand) to thoroughly clean the pus out. Use a band aid or small bandage to cover the boil, and then check it regularly to make sure no other infections occur.

Fungal Infections
The three keys to treating these are: clean, dry, and sunlight. Keep the area as clean and dry as possible, and give it as

much expose to direct sunlight as you can. One very important thing to remember is: *Do not scratch* at the infection. If you have some antifungal powders in your first aid kit, you can use those, but their effectiveness varies.

Rashes

Rashes are generally caused by a reaction to something - like poison ivy, a chemical, or something in the water (or other aspects of the environment). If you can determine what's causing the rash, you can sometimes cure it by merely eliminating that element from your daily life. If that's not possible, you can at least use the following rules:

- If the rash is moist, dry it off and keep it dry
- If the rash is dry, use compresses to keep it moist
- Do not scratch or dig at the rash
- For a wet rash, boil acorns or bark from a hardwood tree to get tannic acid, and then blend it with vinegar. After that, put them on a compress for the area of skin.
- For a dry rash, use some grease or animal fat to rub on the area, and it'll keep it moist

Treat a rash like an open wound; clean and dress it every day. You can also use a number of substances to treat a rash (or wound), and many are available in the wilderness. Here are some of them:

- *Iodine tablets:* put from 5 to 15 tablets of them in a liter (quart) of water to create a rinse that you can apply to a wound or rash while they're healing
- *Garlic:* take a clove of it, slice it open, and then rub it on a boil or wound to spread the oils across the affected area. After that, rinse with clean water
- *Salt water:* while simple, it does kill bacteria. Put 2 to 3 tablespoons in a liter of water, and then apply it to the area
- *Bee's honey:* put it straight on the wound, or dissolve some in water
- *Sphagnum moss:* this moss is found in boggy areas around the world, and is a natural source of iodine. You can put it on a rash or wound like a dressing, and change it daily

Frostbite
This is basically when your tissues become frozen. A mild case results in your skin becoming a dull, whitish color. In more severe cases the frostbite goes deep below the skin and your tissue become solid and you're unable to move it. The vulnerable areas of the body are the hands, feet, and facial areas.

A good way of preventing frostbite is the buddy system. When you're outside in extreme conditions with others, team up with another person, and then each of you check each other's face every once in a while to insure you stay safe. If you're on your own, cover the lower part of your face and nose with your gloves/mittens every once in a while.

If a part of your body is affected by frostbite, never put it near an open flame! Instead, get some lukewarm water in a bowl, sink or tub, place that part of your body in the water, and gently rub it. Afterwards, dry off, and then put the area next to your own skin; this will gradually warm it up. In the event the frostbite is particularly severe, the affected tissue may die. In that case, one of two things will happen:

1. Fingers, toes, ears, even a limb: the extremity will fall off! Bandage the stump and treat as a wound.
2. Tissue: if an area of the body (the calf of the leg, the upper arm etc) is affected, you will essentially have an open wound in the arm or leg. Treats it as such (as outlined above).

Trench Foot
This is an unpleasant condition caused by hours (even days) of being exposed to damp or wet conditions, and with the temperature slightly above freezing. The muscles and nerves in the feet can be damaged, and gangrene can set in. If it becomes extreme, the flesh can die, and then you may have to amputate the foot or even the entire lower leg! To prevent it, keep your feet dry. If you're going to be working outside in cold wet weather, bring extra socks, and change your socks frequently. As soon as you get inside, wash your feet (in fact, wash them daily) and put dry socks on.

Burns
Burns come in four levels, each more severe than the previous, and requiring different treatment. Here are the types of burns you may have to deal with:

1. *First Degree:* like a sun burn, the skin gets red, swells a bit and even peels, and there is pain. Apply lotion, even butter, and/or cool water. Cover with a clean cloth or non-adhesive bandage, and allow to heal.
2. *Second Degree:* the skin and underlying tissue are damaged. The symptoms are similar to First Degree, just more severe.
3. *Third Degree:* the flesh is severely damaged and the patient is going into shock. The flesh can be white or black (charred).
4. *Fourth Degree:* extremely severe, the flesh is burned right down to the deep tissue and/or bones.

Other than First Degree, here's how to treat burns and relieve some of the pain:

- If the subject is still burning, put them out by taking their burning clothes off, throwing water or sand on them, or roll them on the ground.
- Use ice or cold water to cool the burned area.
- Make up some dressings and soak them for 10 minutes in a boiling solution of tannic acid. You can make that from tea, acorns, or the inner bark of hardwood trees.
- Remove the dressings from the solution, cool them, and place them over the burns. If the patient has clothing stuck in the burns - do not remove them, unless they come out easily.
- Treat the burns as you would an open wound.
- Give the subject water to replace fluid loss.
- Make sure their airway is maintained.
- Treat them for shock.
- If you have any painkillers, give them some, unless the burns are on or near the face.

Environmental Injuries
There are a number of health problems that can arise because of your environment and having to live and work under primitive conditions. Here are the major ones, and how to deal with them.

Heatstroke

This is a simple, basic problem you'll encounter, especially if water is in short supply, and it can happen even when the weather is not very hot. The body's regulatory system for temperature breaks down. Watch for these symptoms:

- A swollen, very red face
- The whites of the eyes turning red
- The subject stops sweating
- The subject passes out or becomes delirious
- Bluish color to the lips and nail beds
- The skin grows cool skin

Once the subject reaches those final symptoms, they will be in severe shock. You need to cool them as quickly as possible. If a cool stream or pond is nearby, place them in it, being sure to keep their head above water and their airway open. If you don't have access to one of those, use water (even urine), or cool wet compresses and apply them to all of their joints. Pay special attention to their neck, crotch and armpits, and also wet their head down. If you have an IV, set one up, and then get the person to drink plenty of fluids - water that has a small amount of salt in it is best. Even fanning the person will help.

During the cooling process you can expect the following:

- Vomiting
- Diarrhea
- Struggling, as the person may be delirious. If necessary, hold them down
- Shivering
- Shouting, again, from the delirium. Do your best to calm the person
- Prolonged unconsciousness
- Another bout of heatstroke within the next 2 days
- Cardiac arrest. You may need to perform CPR

Hypothermia

This occurs in cold climates and is when the body is unable to maintain its proper temperature. It differs from heatstroke in that you need to gradually warm the patient up. Have them put on warm dry clothes, give them fluids (warm, but not too hot), and sit them by a fire to slowly warm up. People with hypothermia become sleepy and lethargic. Keep an eye on them and be sure their airway stays open. Do *not* give them alcohol - it cools the body!

Diarrhea

This is a very common problem, and normally isn't a major issue. However, under survival conditions, you don't want some debilitating ailment to weaken you or a family member. It's often caused by changes to your food and/or water, contaminated water, spoiled food, fatigue, dirty dishes, and stress. Given the living conditions you might find yourself contending with, stress will definitely be an issue.

Here are some effective treatments:

- Reduce the intake of fluids for the person for 24 hours
- Drink a cup of strong tea each 2 hours until the diarrhea stops or at least slows. Tannic acid, which is in the tea, helps treat the diarrhea. By boiling the inner bark of any hardwood tree for about two hours or so you can release tannic acid
- Combine a handful of ground up chalk, charcoal or dried bone, and some clean water. If you have the rinds from some citrus fruit, add a portion, it makes the mixture more effective. Give the person two tablespoons every two hours until slowing or stopping the diarrhea

Intestinal Parasites

All manner of worms and intestinal parasites can be avoided, if you take proper measures. Do the following:

- Always wear footwear
- Cook meat and raw vegetables thoroughly
- Don't use raw sewage or human waste as a fertilizer, if possible. If you must, then be sure all crops grown using such materials are washed and cooked.

If you or a loved one gets infected, and you don't have proper medicine, there are home remedies. Here are some you can use:

- *Salt water:* put 4 tablespoons of salt in a liter (quart) of water and then drink it. However, this is a one-time treatment - do not repeat it as salt water, in large quantities, is harmful

- *Tobacco:* take a cigarette and eat it! While distasteful, the nicotine will kill, or at least stun, the worms, and then the person can pass them. If they have a severe infestation, repeat the treatment in 1 to 2 days, but do *not* do it sooner
- *Kerosene:* give the patient 2 tablespoons to drink, *but no more.* You can repeat this 1 to 2 days, but be careful not to breathe in the fumes
- *Hot peppers:* these are an effective means of keeping parasites out of your body, but only if eaten regularly. They can be eaten raw or in soups, and with rice and meat dishes

Herbal Remedies
When living off the land, making use of natural plants to treat various problems will be something you must get used to. With that in mind, here are some remedies you can make use of. First, here are some terms and definitions associated with plants used for medicine:

- *Poultice:* crushed leaves or plant parts, sometimes heated, and then applied directly or in cloth/paper to a wound or a sore
- *Infusion or tisane or tea:* prepared by putting a small amount of an herb in a container, adding hot water, and letting it steep before you use it
- *Decoction:* an extract made by adding an herb leaf or root to water and bringing it to a continuous boil or simmer it to draw the chemicals out and into the water
- *Expressed juice:* this is the sap or liquid from a plant, squeezed out and then applied to a wound or used to make a medicine
- Here are some remedies you can use:
- *Diarrhea:* use blackberry roots (and similar berries) to make a tea. White oak bark, and similar trees, have tannin, and this is also effective. Eating the ashes of a campfire or white clay also works. Teas made from cranberry, cowberry, and hazel leaves work too
- *Antihemorrhagics:* to help stop bleeding, make a poultice from the puffball mushroom, plantain leaves, or (the most effective) yarrow or woundwort leaves
- *Antiseptics:* use juice from wild onions, garlic chickweed leaves, or the crushed leaves of dock. This juice is useful for cleaning wounds, rashes, and sores. You can

also make a decoction from burdock root, the roots or leaves of the mallow, or white oak bark. Only use these externally

- *Fevers:* make tea out of willow bark. You can also make an infusion from elder flowers or fruit, or tea from linden flowers, or a decoction from elm bark
- *Colds and sore throats:* make a decoction from plantain leaves or willow bark. Use mallow, burdock roots, or mullein roots or flowers, or mint leaves to make tea
- *Aches, pains, and sprains:* apply an external poultice made from dock, chickweed, plantain, garlic, willow bark, or sorrel. There are salves you can make by combining the expressed juices of those same plants with vegetable oils or animal fats
- *Itching:* make a poultice out of jewelweed or witch hazel leaves
- *Sedatives:* make tea out of mint leaves or the leaves of the passionflower
- *Hemorrhoids:* make an external wash from oak bark or elm bark tea, or from the expressed juice of leaves from the plantain, or a decoction of Solomon's seal root
- *Constipation:* make decoctions out of dandelion leaves, walnut bark, or rose hips, and then drink it. If you eat raw daylily flowers they'll also work
- *Gas and cramps:* make tea from carrot seeds or mint leaves
- *Antifungal washes:* use walnut leaves, oak bark, or acorns to make a decoction. Apply frequently, and alternate with exposing the affected area to direct sunlight

Marijuana
This may sound like an odd plant to grow for medical uses, but it is effective as a painkiller and treatment for certain eye problems. If the crisis you're dealing with is truly a global breakdown of society, then concerns about law enforcement will be the least of your worries. You can smoke it, eat it, and use it as an ingredient in foods. In addition, it can help you to "chill out" from the stress of dealing with the breakdown of society - and the loss of all that you've known! However, it is critical that you not over-indulge in it, as it can become addictive. Remember, you need to stay focused on survival.

Finally, like alcohol, cigarettes, and other products, you can use marijuana to trade and barter with other survivors.

SIGNALING TECHNIQUES

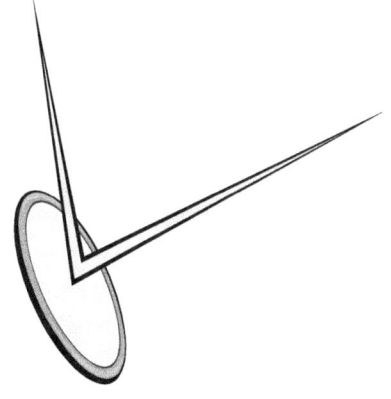

When you and your loved ones are out in the wilderness, there is going to come a time when you'll want to contact others. The disaster will be over, society will start to re-build, and you're going to want to get in touch with rescue parties and/or other survivors. Or, in the worst-case scenario, you and your family and other people who have likewise taken shelter will be among the only survivors. In that case, you'll want to make contact with such people and see about creating a new society.

Either way, a major issue for you will be summed up in one word: trust. Who can you trust; which people are going to be friendly and peaceful? To find this out, you need to make contact with these people, yet keep them at a distance until you get to know them.

This is where different types of signaling come into play. The first and best means of communication is - of course - a phone and/or radio. The problem with these types is simple: power. In the case of a cellphone, there's also the issue of the cell towers. If the crisis that has befallen the world is truly global and catastrophic in nature, those towers will probably be inactive. In the case of a shortwave radio, not only is there the matter of power, but the other people will have to have a radio, and they'll have to be on the same frequency as you.

All of these circumstances are highly unlikely!

So, your best means of signaling others is with a simpler method. Let's look at some.

Fire
A very simple and basic means of communication; this is highly effective at night, which is an added plus to its use. When you're first trying to contact others, lighting a signal fire on a

high promontory - far from your safe house - at night and just letting it burn will let others in the area know that you're there. Then, each night, look for others to reply in kind. Once you've established communication, you can use a sheet or blanket to block the fire, and thus send messages via Morse code.

You can easily find Morse code online or in a book, and then write it down. The most basic message is a distress signal consisting of S-O-S: three dots, three dashes, and then three dots. Another means of signaling you're in need of assistance is to build three fires in a triangular arrangement.

If you're in a jungle environment, try to find a large open and clear area, or use some equipment to clear an area. You do this for two reasons: you want good visibility, and you do not want to chance setting the foliage on fire. A raging forest fire is the last thing you want to deal with! If the area has snow or ice, build a small platform or deck for the fire to keep it high and dry. Otherwise the fire will tend to be put out by the melting snow.

Smoke
This is the best signaling technique in the daylight. A large column of black smoke can easily be seen for miles, on a clear day. To create smoke, stoke up a good fire, and then dump moss, green leaves, and even a little water on it. White smoke will be created. If you have some pieces of rubber or can dose some rags with oil, tossing them on the fire will make black smoke. As with fire, three columns of smoke are the international signal that you're in distress and need help.

In the event you've received word that conditions are improving in the world - maybe from the radio or neighbors - and that the government is launching rescue efforts, a fire or smoke distress signal will be ideal for getting help.

Beyond smoke and fire, there are a plethora of devices you can use to signal others. Here's a breakdown of them:

- Smoke Grenades: you can generally get them from an army surplus store or camping equipment store. As they are very hot when they ignite, be sure to only toss them into an open area free of vegetation.

- Pen Flares: these are typically given to pilots, so you may not have access to them. They - as their name implies - look like a pen. You hold them above your head, pull the cord, and they shoot off a flare with a loud bang. They have the virtue of being small; you can easily wear one around your neck or carry it in your pocket. That way, if you hear a plane approaching, you can shoot one off.
- Tracer Ammunition: these are bullets that create a streak of light when fired. To signal someone, shoot these rounds into the air, *not* at the person or vehicle you're trying to signal.
- Star Clusters: yet another type of flare; they come in different colors, and red is the color for international distress. So, you can shoot off a red one to signal a rescue party.
- Star Parachute Flares: these are similar to other flares, but have the virtue of burning longer. When you shoot one off, it deploys a small parachute that slows its descent, and thus people can see it longer.

There are also a number of very simple methods for signaling people/rescue parties. You can make use of items from your safe house. Here's a breakdown of what to use and how:

- Mirror/Shiny Object: so long as it is a clear day and the sun is shining, you can use a mirror or other reflective item to signal others. As with a fire, use Morse code to relay messages. There is also a specially designed mirror made just for messaging. It has a small clear spot at its center. You hold the mirror to your face, look through the opening, and then turn the mirror until you see the sunlight reflected on the plane, vehicle etc that you're trying to signal. By gently turning the mirror, the light will appear to them to flash, and thus communication can be established.
- Flashlight: this works best at night, and as with other methods for signaling, you can use Morse code to send and receive messages.
- Clothes: while this may seem silly, it is effective. Pick out brightly colored items and either spread them on the ground or hang them high in nearby trees.

- Natural Items: you can cut down trees and arrange the trunks in a pattern to create words - SOS is generally easiest. You can clear shrubs to do the same; arrange rocks and stones, pile sand to form letters, dig trenches, even pile leaves and/or twigs.

Audio Communication

Yet another method for contacting people is with sound. Depending on your budget and location, there are several means of doing this.

Radios

As mentioned before, a radio is the best method. Not only will you be able to contact people very far away, but you can keep updated on the worldwide situation. There are many different brands and models of radios; so we won't go into all of them - that would be a book in itself. Instead, let's consider some general principles to keep in mind when getting a radio:

- When not in use, turn the unit off. You have no way of knowing how long the disaster is going to last; you need to conserve power as much as possible.
- Protect the radio's power source from heat and cold; both are detrimental to batteries.
- Place the antenna as high up as possible. With a standard radio, it operates on what is known as LOS (Line of Sight). That is - the signal goes in a straight line from your antenna to the next one. If large terrain features like hills, bluffs etc are in the way, you won't get a signal through.
- Keep the antenna clean. If anything touches the antenna, such as leaves, tree branches, clothing, and even your skin, the range of your signal will be reduced.

Whistles

While a very simple means of communication, they are also quite effective. You can either buy one from a store or learn to do it yourself. Under the right conditions, a human whistle can carry for upwards of a mile or more; those bought at a store can be heard even further. They have the virtue of being able to be used during the day or night, and you can again use Morse code to send messages.

Gunshots

A very basic means of signaling; you really can't use this for long messages - just a distress call. It is unwise to use live ammunition; if you can, get some blanks and have them on hand for an emergency.

KNOTS & CONSTRUCTION LASHINGS

Various knots are going to be needed for you to secure vital equipment, tie up traps and snares, secure livestock, protect items from the environment (such as being blown away in a storm), and so on. As it happens, there are quite a few types of knots; so we'll review the basics of the main ones.

Square Knot
Perhaps the simplest of all knots; it is made by following four easy steps:

1. Hold one end of each rope in each hand, and put the rope in your right hand over the rope in your left.
2. Pull the rope under and then over the left hand rope.
3. Take the left hand rope, which is now in your right hand, put it over the right hand line, pull it under, and then over.
4. Pull the knot tight.

Figure 41. Square Knot.

Fisherman's Knot

This is especially good for tying two different ropes together, and is accomplished in three easy steps:

1. Tie one end in an overhand knot over a short length of the other line.
2. Tie the other line also in an overhand knot a short distance from the first knot.
3. Slide the knots tightly close to each other and tighten them.

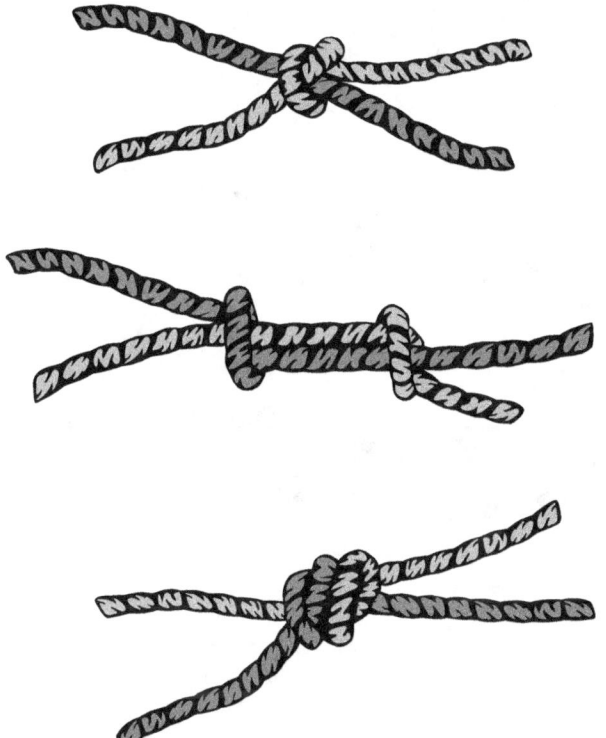

Figure 42. Fisherman's Knot

Double Fisherman's Knot

This is a slight variation of the previous knot, and takes only three steps to complete. Here is how to make one:

1. Tie two wraps of one end of one rope around the other.
2. Repeat the process with the free end of the other rope, tying the knot a short distance from the first knot.
3. Pull the two knots close together, and tighten both of them down.

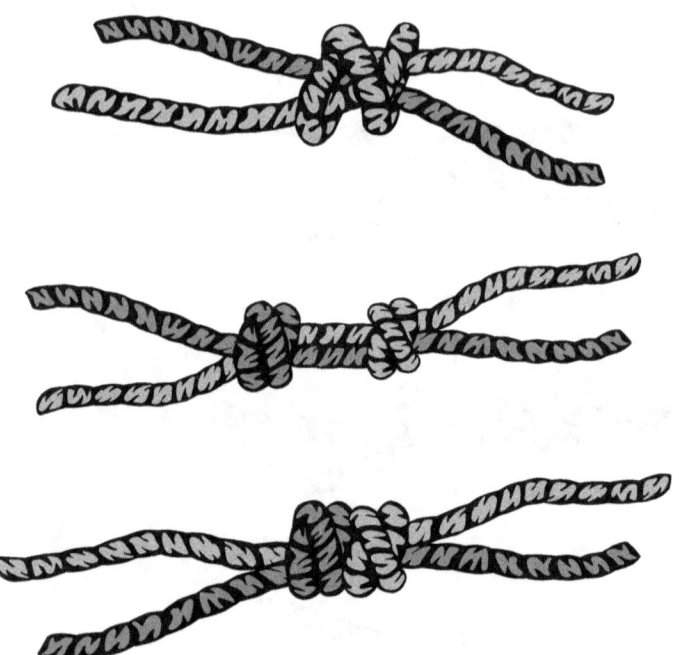

Figure 43. Double Fisherman's Knot.

Bowline Knot

This is a very simple knot to tie, and is also very useful for a myriad of purposes. There are three steps to creating it:

1. Twist the rope to create an eye a short distance from the end.
2. Feed the end through the eye; be sure to leave enough rope so you have a decent sized loop.
3. Loop the end around the rope above the eye, and then bring it back through the eye.
4. Pull the ends tight until the eye is completely closed and the loop is formed.

Figure 44. Bowline Knot.

Figure Eight Retrace

This is a good knot for tying ropes to trees. While it takes several more steps than other knots, it is still easy to carry out, as follows:

1. Measure out enough rope to get around whatever it is you want to tie the rope to.
2. Make a loop in the rope, wrap the end of the rope around the rope, and feed it through the loop. This is a figure-eight knot. Do *not* tighten it, and make sure the knot is far enough from the end of the rope to allow the rope to get around the tree, rock, or whatever it is you're going to tie the rope to.
3. Loop the end of the rope around the object you want to tie to.
4. Slip the end through the figure-eight knot, around the main part of the rope, and then through the figure-eight again.
5. Pull the rope tight.

Figure 45. Figure Eight Retrace

Clove Hitch

This is a good knot to use when you need to tie a rope to a taut cable, rope, or some other sort of thin object. It is made by doing the following:

1. Place the rope on top of the other rope or object you want to tie it to.
2. Feed the rope under the object, and then loop it over.
3. Pass the end of the rope back over the object, but under itself (the first pass of the rope).
4. Do *not* tighten the rope yet! Pass as much of the rope as you feel you need on that end (perhaps to tie to something else), and then tighten the loops down.

Figure 46. Clove Hitch

Wireman's Knot

This is a knot you tie in the middle of a rope to create a loop. It takes several steps to accomplish, and is done as follows:

1. Take the rope and wrap it twice around your left hand (keeping your palm up), and wrap from left to right.
2. Each of the sections of the rope lying across your hand is designated with a name - for ease of following the instructions. They are: fingertips, palm and heel.
3. Take hold of the palm section with the finger and thumb of your right hand, and put it over the heel section.
4. Grab the heel section and put it over the fingertip section.
5. Grab the fingertip and put it over the palm.
6. Take hold of the palm section and gently pull to form the loop.
7. Pull on the loop and the other two ends of the rope to draw the knot tight.

Figure 47. Wireman's Knot.

Directional Figure-Eight

This is another knot you can tie in the middle of a rope. However, it is much easier to tie than the previous knot. These are the steps:

1. Lay the rope in your left palm.
2. Pull up the free end of the rope to make a large loop, and lay the rope in your palm next to the other line.
3. Tie a figure-eight knot in the rope using the end of the loop; be sure to pull the loop through the knot as far as possible - this will insure you have an adequate loop for whatever it is you need.
4. Pull the knot tight.

Figure 48. Directional Figure-Eight.

Figure-Eight Loop

Yet another of the knots you can tie in a rope to form a loop in the middle of the rope, it is quite simple to create. Just follow these steps:

1. Double over the rope for a length long enough to satisfy the size of the loop you need.
2. Form a loop in the rope.
3. Tie a figure-eight knot in the rope, making sure to create a loop big enough for your need.
4. Tighten the knot.

Figure 49. Figure-Eight Loop.

Prusik Knot

This is one of a series of specialty knots, and you can tie it along any point on a rope. It is best used to tie one rope to another, and is made by doing the following:

1. Double the rope up, and lay the end of the loop over the tightened rope you want to tie to.
2. Pass the two ends of the rope through the loop.
3. Pass the ends through the loop again.
4. Tighten the loops down, and then tie an overhand knot near the first knot to keep it tight.

Figure 50. Prusik Knot.

Transport Knot

This is a simple overhand knot, and it can be used to transport objects. You form it by following these steps:

1. Form a loop in the rope.
2. Form a twist in the rope near the loop.
3. Pass the twist over the loop.
4. Pull the twist tight to create the knot; be sure to leave enough of a loop for your needs.

Figure 51. Transport Knot.

Kleimhiest Knot

This is an ideal knot for carrying large loads, when you need to secure the rope to another rope or a cable. Just follow these steps to make it:

1. Find the middle of the rope, and double it over.
2. Take the doubled rope and wrap it about the taut line several times (four is the minimum to be effective and strong).
3. Take the two ends of the rope and pass them through the end of the loop.
4. Pull the wraps together as you also pull the ends of the rope so as to pull the loop tight.
5. Tighten down the entire knot until all of the wraps are snug and tight.

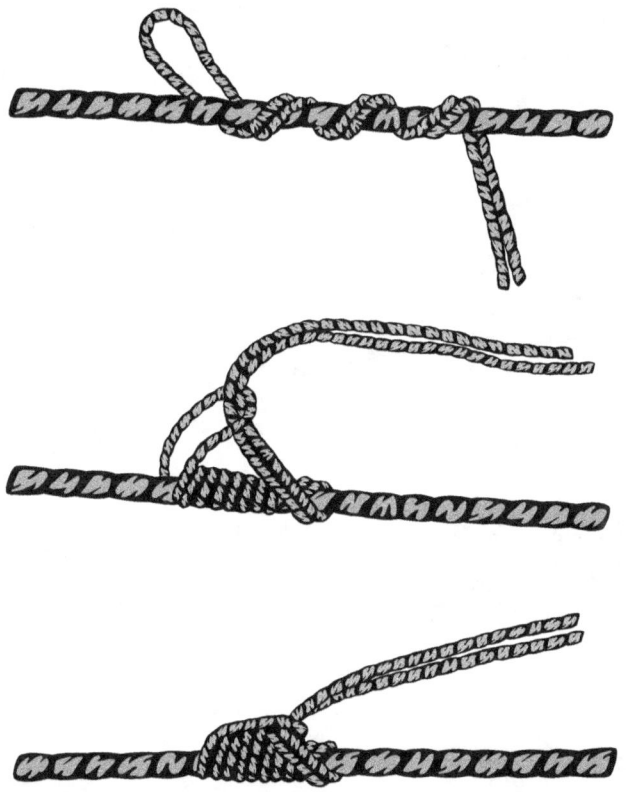

Figure 52. Kleimheist Knot.

The Albright Special

This knot is ideal for tying together two different ropes, especially when they are not of the same diameter and/or material. Here are the steps to creating it:

1. Make a loop in the thicker of the two ropes.
2. Pass the end of the other rope through the loop; make sure you have plenty of length.
3. Wrap the thinner rope around the other rope at least ten times.
4. Pass the thinner rope back through the loop; make sure the rope goes out the loop the same way it went in.
5. Hold both ends of both of the ropes and pull on them; slowly tightening down the loop and wrap until the knot is tight.

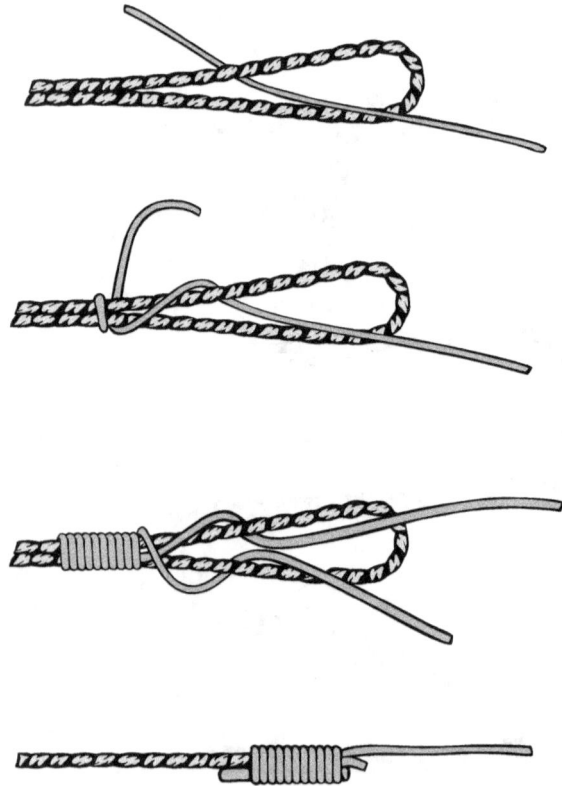

Figure 53. The Albright Special

Surgeon's Knot

Not to be confused with the knot a surgeon ties during an operation; this knot is very strong and can be tied very quickly. It is also good for connecting ropes of different diameters together. Here are the steps in making one:

1. Lay the two lines next to each other so that they overlap for a short length.
2. Form a simple loop.
3. Put the end of the rope from the reel (as indicated on the diagram below), and all of the leader rope twice through the loop.
4. Pull on all four ends of the ropes until the knot is tightened.

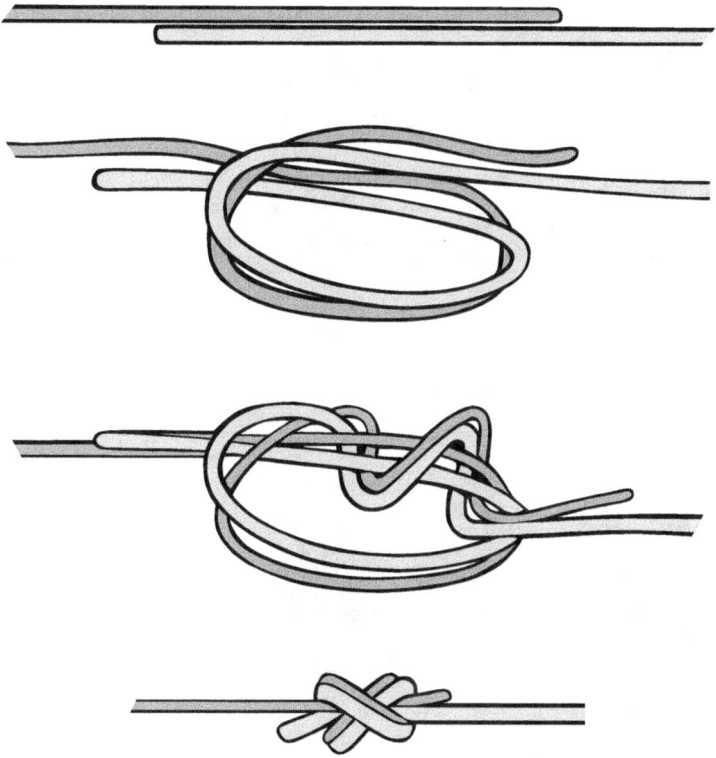

Figure 54. Surgeon's Knot.

Perfection Loop

This is a very compact and neat knot that allows you to form a loop at the end of a rope. The steps to make it are as follows:

1. Make a small loop in the end of the rope, making sure to leave enough rope at the end for a second loop and a little extra.
2. Use the end of the rope to make a second loop, and hold the two loops so that they form roughly a right angle.
3. Lay the end of the rope between the two loops.
4. Take the second loop and pass it through the first one.
5. Pull on the second loop until the knot is tight.

Optional: you can trim the end of the rope so it doesn't get in the way.

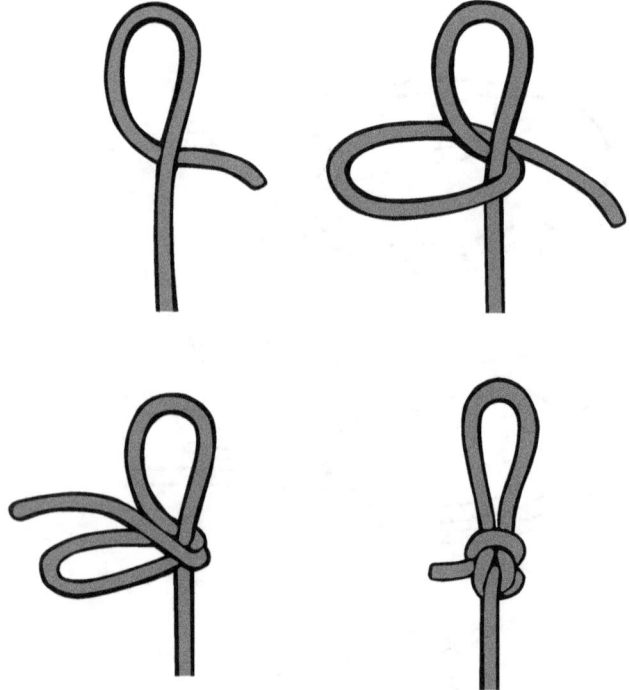

Figure 55. Perfection Loop.

Sheepshank

This is a good knot to use if you have a rope with a frayed section to it and you're worried about it breaking. By tying a sheepshank, you can strengthen the rope. Also, if you want to shorten a rope, but not cut it, this is an ideal knot to use. Here are the steps to creating it:

1. Pull up some slack in the rope and make two loops (A and B, see below).
2. Take the small bights (A1 and B1), and pass them through loops A and B, respectively.
3. Pull on both ends up the rope to tighten the knot.

Figure 56. Sheepshank Knot.

www.ingramcontent.com/pod-product-compliance
Lightning Source LLC
Chambersburg PA
CBHW070137290526
45789CB00002B/515